Synthesis, Characterization, and Applications of Functional Materials—Thin Films and Nanostructures

MATERIALS RESEARCH SOCIETY
SYMPOSIUM PROCEEDINGS VOLUME 1675

Synthesis, Characterization, and Applications of Functional Materials—Thin Films and Nanostructures

Symposium held April 21-25, 2014, San Francisco, California, USA

EDITORS

Valentin Craciun
National Institute for Laser, Plasma,
and Radiation Physics
Magurele, Ilfov, Romania

Maryline Guilloux-Viry
Institute of Chemical Sciences ISCR
Rennes, France

Menka Jain
University of Connecticut
Storrs, Connecticut, USA

Quanxi Jia
Los Alamos National Laboratory
Los Alamos, New Mexico, USA

Hiromitsu Kozuka
Kansai University
Suita, Japan

Dhananjay Kumar
North Carolina A&T State University
Greensboro, North Carolina, USA

Sanjay Mathur
University of Cologne
Cologne, Germany

Xavier Obradors
Institut de Ciencia de Materials
de Barcelona
Bellaterra, Catalunya, Spain

Kaushal K. Singh
Applied Materials
Santa Clara, California, USA

Materials Research Society
Warrendale, Pennsylvania

CAMBRIDGE
UNIVERSITY PRESS

Shaftesbury Road, Cambridge CB2 8EA, United Kingdom

One Liberty Plaza, 20th Floor, New York, NY 10006, USA

477 Williamstown Road, Port Melbourne, VIC 3207, Australia

314–321, 3rd Floor, Plot 3, Splendor Forum, Jasola District Centre, New Delhi – 110025, India

103 Penang Road, #05–06/07, Visioncrest Commercial, Singapore 238467

Cambridge University Press is part of Cambridge University Press & Assessment,
a department of the University of Cambridge.

We share the University's mission to contribute to society through the pursuit of
education, learning and research at the highest international levels of excellence.

www.cambridge.org
Information on this title: www.cambridge.org/9781605116525

© Materials Research Society 2014

First published 2014

CODEN: MRSPDH

A catalogue record for this publication is available from the British Library

ISBN   978-1-605-11652-5   Hardback

CONTENTS

ZnO THIN FILMS AND NANOSTRUCTURES

MULTIFERROICS, MAGNETISM, AND MAGNETIC MATERIALS

OXIDE THIN FILMS AND NANOSTRUCTURES

PREFACE

Symposium K, "Nanostructures, Thin Films and Bulk Oxides—Synthesis, Characterization and Applications" and Symposium RR, "Solution Synthesis of Inorganic Functional Materials" were held April 21–25 at the 2014 MRS Spring Meeting in San Francisco, California.

Oxide materials from bulk down to nanostructures are used for applications in microelectronics, communications, sensing, energy, catalysis, nanophotonics, and optoelectronics. As the characteristic dimensions of oxide systems shrink into the nanometer range, there are increased technological challenges for synthesis, processing, and characterization to ensure high uniformity, reproducibility, and cost reduction.

This symposium proceedings volume represents the recent advances in various areas of deposition, processing, characterization, and integration of functional oxide materials, with particular emphasis on the relationship among the structure, composition, stability and functional properties. The papers are divided into three sections: (1) ZnO Thin Films and Nanostructures, (2) Multiferroics, Magnetism, and Magnetic Materials and (3) Oxide Thin Films and Nanostructures. The papers published in this volume provide answers to many scientific questions regarding the role of interfaces, defects, composition, stress and size effects on their properties and functionalities and offer insight into the exciting recent developments occurring in oxide materials from bulk down to nanostructures. We hope that the volume is a valuable tool in guiding and informing the scientific community about new and important advancements happening in the area of oxide materials.

Valentin Craciun
Maryline Guilloux-Viry
Menka Jain
Quanxi Jia
Hiromitsu Kozuka
Dhananjay Kumar
Sanjay Mathur
Xavier Obradors
Kaushal K. Singh

October 2014

Acknowledgments

The papers published in this volume result from two MRS Spring 2014 symposia—K and RR. We sincerely thank all of the oral and poster presenters of the symposia who contributed to this proceedings volume. We also thank the reviewers of these manuscripts for their work and valuable feedback to the editors and authors. It is an understatement to say that the symposia and the proceedings would not have happened without the organizational help of the Materials Research Society.

MATERIALS RESEARCH SOCIETY SYMPOSIUM PROCEEDINGS

MATERIALS RESEARCH SOCIETY SYMPOSIUM PROCEEDINGS

Prior Materials Research Symposium Proceedings available by contacting Materials Research Society

ZnO Thin Films and Nanostructures

Mater. Res. Soc. Symp. Proc. Vol. 1675 © 2014 Materials Research Society
DOI: 10.1557/opl.2014.860

Optimization of Annealing Conditions for ZnO-based Thin Films Grown Using MOCVD

Anas Mazady, Abdiel Rivera, and Mehdi Anwar

Electrical and Computer Engineering, University of Connecticut, Storrs, CT 06269, Email: anwara@engr.uconn.edu

ABSTRACT

In this work, effects of thermal annealing on the structural and optical properties of ZnO thin films grown on p-Si and GaN substrates using metalorganic chemical vapor deposition (MOCVD) are investigated. Annealing at 600 °C results in optimum crystal and optical qualities of the ZnO thin films on both substrates. Smaller lattice mismatch between grown ZnO epitaxial layer on GaN substrates results in better film morphology as compared to p-Si substrates. Higher annealing temperature along with a slower thermal ramp provides better crystal quality of ZnO thin films on both substrates. Annealing ZnO thin films at 700 °C with a slower thermal ramp results in better crystal quality as is evident from a 56% reduction in the full-width at half maximum (FWHM) of the (002) peak compared to the as-grown films. The optical quality also enhances with a slower annealing rate. The determination of the optimum annealing conditions for different substrates has important implications in fabricating optimized and efficient ZnO based electronics.

INTRODUCTION

ZnO, being a transparent conductive oxide, has drawn considerable interests in fabricating transparent electrodes for display devices, transparent electronics, and photovoltaic devices, to name a few [1]. A large direct bandgap energy of 3.37 eV makes it an ideal candidate for optoelectronic applications, such as, solar blind ultra-violet (UV) detectors, UV diodes, LEDs, among others. Reasonable smaller lattice mismatch of 2% with GaN also allows ZnO to be a good substrate material for GaN based devices, as stand-alone GaN substrates are not available yet [2]. Being resistant to radiation damage, compared with other semiconductors, makes ZnO a suitable candidate for space applications [3].

The large lattice mismatch of 40% and large difference in the thermal expansion coefficients of 87% between ZnO and Si substrates cause built-in residual stress in the grown ZnO NWs. Proper annealing conditions can reduce this built-in residual stress and hence can improve the crystalline quality and minimize defects of the grown ZnO NWs. Post annealing not only passivates native defects but also improves the near band-edge emission of ZnO films through reduction of non-radiative recombination centers in the films [4]. It also produced a smooth surface of the film required for most optoelectronic application. Sengupta et al. [5] investigated the annealing effects on ZnO film grown via sol-gel. They observed that the intensity of (002) peak gradually increases with increasing annealing temperature, giving rise to a smaller full width at half maxima (FWHM) and an average larger grain size. This phenomenon is attributed to the thermal annealing induced coalescence of small grains by grain boundary diffusion, resulting in larger grain size [6].

In this work, effects of thermal annealing on the structural and optical properties of ZnO thin films grown using metalorganic chemical vapor deposition (MOCVD) are investigated. Mahmood et al. [7] performed similar studies on ZnO thin films deposited using reactive e-beam evaporation technique. However, material quality can be significantly improved by employing MOCVD which is also a standard technique for mass production in the industries.

EXPERIMENT

ZnO thin films were deposited on p-Si and GaN substrates using FirstNano EasyTube 3000 MOCVD system at a constant temperature and pressure of 300 °C and 70 Torr, respectively. Diethylzinc (DEZn) and N_2O were used as the zinc and oxygen precursors and nitrogen was used as the carrier gas. The growth was carried out for 20 mins maintaining a steady flow of 50 SCCM and 35 SCCM of DEZn and N_2O, respectively. After deposition of the thin film, the samples were annealed at 500 – 750 °C under N_2 ambient.

DISCUSSION

The scanning electron microscope (SEM) images of ZnO thin films grown on p-Si substrates are shown in Figure 1. The as-grown film on p-Si substrate shows granular surface texture with very fine grains. Thermal annealing treatments at 500 and 600 °C result in a smoother film with smaller RMS roughness. For annealing temperatures higher than 700 °C, large grains were visible with the grain size varying between 35 and 155 nm. In contrast, ZnO thin films grown on GaN substrates showed better morphology in general, as shown in Figure 2, which is attributed to a smaller lattice mismatch between ZnO and GaN. For comparison, the lattice mismatch exerts a 40% tensile strain on ZnO thin films grown on p-Si substrates, whereas the strain is only 2% (compressive) in the case of ZnO thin films grown on GaN substrates. Annealing at temperatures higher than 600 °C resulted in a very smooth surface of the film on GaN substrates.

Figure 1. SEM images of ZnO (thin film) on p-Si substrates

Figure 2. SEM images of ZnO (thin film) on GaN substrates

In order to investigate the origin of the granular textures, the samples were annealed and cooled at a slower rate of 2.5 °C/min which is twice as slower than the regular annealing condition used in this study. SEM images of the thin film annealed at different rates and temperatures are shown in Figure 3. The slower annealing/ cooling rate did not show any observable difference in the grain size of ZnO thin films on different substrates compared to their faster annealed counterparts. This suggests that the observed granular surface texture at higher annealing temperature is not a direct result of the thermal coefficient mismatch. Rather, at high annealing temperatures the grain boundaries migrate and the adjacent grains coalesce, independent of the annealing/ cooling ramp [8].

Figure 3. Slow Vs. fast annealing/cooling ramp for ZnO thin film grown on p-Si and GaN substrates

Figure 4 shows X-ray diffraction (XRD) measurement results of ZnO thin films of p-Si substrates. The as-grown film has multiple growth facets, along (100), (002), and (200). The film became more single crystalline with (002) orientation, when annealed at 600 °C. It is also important to note that, although a slow annealing does not affect the morphology, it improves the crystalline quality significantly. It is clear from the figure that the (002) peak has a much higher intensity than the other peaks for the samples annealed/ cooled at a slower thermal ramp than those at a faster ramp.

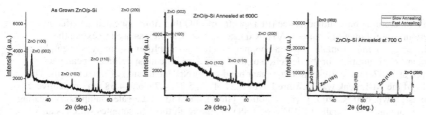

Figure 4. XRD measurement results of ZnO thin film grown on p-Si substrates

Table 1. Summary of crystalline quality of ZnO thin film on p-Si substrates annealed at different temperatures

	As-Grown	500 °C	600 °C	700 °C Slow	700 °C Fast
c-lattice (Å)	5.1921	5.1906	5.1820	5.1870	5.1891
a-lattice (Å)	3.2430	3.2430	3.2406	3.1792	3.2417
Misfit Strain	67.47%	67.47%	67.59%	70.83%	67.53%
FWHM (deg)	0.2225	0.1279	0.1280	0.09695	0.1239

Table 1 summarizes the crystalline quality of ZnO thin films grown on p-Si substrates and annealed at different temperatures. The c-lattice constant is observed to decrease with increased annealing temperature which is in agreement with the observations made by other groups [9-11]. Comparison of the full width at half maximum (FWHM) of the (002) peak suggests that the crystalline quality becomes better with higher temperature annealing and a slower anneal/ cooling ramp increases the film quality to a higher degree. For an example, the 700 °C slow annealed sample has a 56% smaller FWHM than the as-grown film.

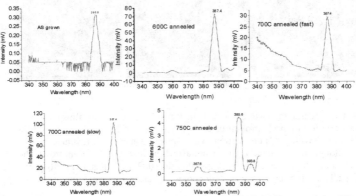

Figure 5. Room temperature photoluminescence (PL) spectra of ZnO/p-Si thin films annealed at different temperatures

Room temperature photoluminescence (PL) spectra of ZnO thin films annealed at different temperatures are shown in Figure 5. A He-Cd laser line at 325 nm was used to excite the samples, and a monochromator, optical chopper, and lock-in amplifier assembly was employed in standard configuration for the detection of the PL spectra. The most dominant peak was observed at ~387.4 nm corresponding to near band-edge (NBE) emission of free excitons. The FWHM of the NBE peak was measured to be 5.21, 4.38, 5.21, 4.79, and 4.38 nm, respectively, for the as-grown, 600 °C, 700 °C (fast), 700 °C (slow), and 750 °C annealed samples. Annealing at 600 °C obtained the smallest FWHM of 4.38 nm among all the p-Si samples which is also 16% smaller than the as-grown p-Si samples. The 750 °C annealed sample also had a 4.38°

FWHM of the NBE peak, but another peak at 358 nm emerges, which is due to the band to band transition of electrons [12].

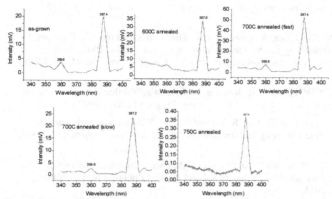

Figure 6. Room temperature photoluminescence (PL) spectra of ZnO/GaN thin films annealed at different temperatures

PL spectra of ZnO thin films grown on GaN substrates and annealed at different temperatures are shown in **Error! Reference source not found.**. FWHM of ZnO/GaN thin films are determined to be 4.58, 4.58, 4.79, 4.58, and 4.79 nm, respectively. The peak intensity of the 600 °C annealed ZnO/GaN sample was 69% higher than the as-grown film. All the measurements therefore suggest 600 °C to be the optimum annealing temperature for ZnO thin films grown on both p-Si and GaN substrates.

CONCLUSIONS

ZnO thin films were grown on p-Si and GaN substrates using MOCVD. The films were annealed at 500, 600, 700, and 750 °C at different anneal/ cooling rates under N_2 ambient, *in situ* in the process chamber. Least surface roughness of the ZnO/p-Si thin films was obtained when annealed at 500-600 °C window. GaN substrates obtain better morphology of ZnO thin films than p-Si substrates due to a smaller lattice mismatch. Higher annealing temperature with a slower thermal ramp obtains better crystal quality of ZnO thin films on both substrates. Optical quality also enhances for a slower annealing rate.

ACKNOWLEDGMENTS

The research was partially supported by NSF I/UCRC, Magnolia Optical Technologies, Inc., NAVAIR, Department of Education, and the Center for Hardware Assurance, Security, and Engineering (CHASE). The authors would like to acknowledge the support of Drs. Ashok Sood, Tariq Manzur and John Zeller.

REFERENCES

[1] X. Jiang, F. Wong, M. Fung and S. Lee, "Aluminum-doped zinc oxide films as transparent conductive electrode for organic light-emitting devices," *Appl. Phys. Lett.,* vol. 83, pp. 1875-1877, 2003.

[2] Z. Chen, S. Yamamoto, M. Maekawa, A. Kawasuso, X. Yuan and T. Sekiguchi, "Postgrowth annealing of defects in ZnO studied by positron annihilation, x-ray diffraction, Rutherford backscattering, cathodoluminescence, and Hall measurements," *J. Appl. Phys.,* vol. 94, pp. 4807-4812, 2003.

[3] D. C. Look, D. Reynolds, J. W. Hemsky, R. Jones and J. Sizelove, "Production and annealing of electron irradiation damage in ZnO," *Appl. Phys. Lett.,* vol. 75, pp. 811-813, 1999.

[4] Y. Wang, S. Lau, X. Zhang, H. Hng, H. Lee, S. Yu and B. Tay, "Enhancement of near-band-edge photoluminescence from ZnO films by face-to-face annealing," *J. Cryst. Growth,* vol. 259, pp. 335-342, 2003.

[5] J. Sengupta, R. Sahoo, K. Bardhan and C. Mukherjee, "Influence of annealing temperature on the structural, topographical and optical properties of sol–gel derived ZnO thin films," *Mater Lett,* vol. 65, pp. 2572-2574, 2011.

[6] Y. Caglar, S. Ilican, M. Caglar, F. Yakuphanoglu, J. Wu, K. Gao, P. Lu and D. Xue, "Influence of heat treatment on the nanocrystalline structure of ZnO film deposited on p-Si," *J. Alloys Compounds,* vol. 481, pp. 885-889, 2009.

[7] A. Mahmood, N. Ahmed, Q. Raza, T. M. Khan, M. Mehmood, M. Hassan and N. Mahmood, "Effect of thermal annealing on the structural and optical properties of ZnO thin films deposited by the reactive e-beam evaporation technique," *Phys. Scripta,* vol. 82, pp. 065801, 2010.

[8] Y. Lin, J. Xie, H. Wang, Y. Li, C. Chavez, S. Lee, S. Foltyn, S. Crooker, A. Burrell and T. McCleskey, "Green luminescent zinc oxide films prepared by polymer-assisted deposition with rapid thermal process," *Thin Solid Films,* vol. 492, pp. 101-104, 2005.

[9] V. Gupta and A. Mansingh, "Influence of postdeposition annealing on the structural and optical properties of sputtered zinc oxide film," *J. Appl. Phys.,* vol. 80, pp. 1063-1073, 1996.

[10] M. Puchert, P. Timbrell and R. Lamb, "Postdeposition annealing of radio frequency magnetron sputtered ZnO films," *Journal of Vacuum Science & Technology A,* vol. 14, pp. 2220-2230, 1996.

[11] Y. Lee, S. Hu, W. Water, K. Tiong, Z. Feng, Y. Chen, J. Huang, J. Lee, C. Huang and J. Shen, "Rapid thermal annealing effects on the structural and optical properties of ZnO films deposited on Si substrates," *J Lumin,* vol. 129, pp. 148-152, 2009.

[12] G. Lee, Y. Yamamoto, M. Kourogi and M. Ohtsu, "Blue shift in room temperature photoluminescence from photo-chemical vapor deposited ZnO films," *Thin Solid Films,* vol. 386, pp. 117-120, 2001.

Mater. Res. Soc. Symp. Proc. Vol. 1675 © 2014 Materials Research Society
DOI: 10.1557/opl.2014.886

ZnO Nanostructures on Electrospun Nanofibers by Atomic Layer Deposition/Hydrothermal Growth and Their Photocatalytic Activity

Fatma Kayaci,[1,2] Sesha Vempati,[*1] Cagla Ozgit-Akgun,[1,2] Necmi Biyikli[1,2] and Tamer Uyar[**1,2]
[1] UNAM-National Nanotechnology Research Center, Bilkent University, Ankara, 06800, Turkey
[2] Institute of Materials Science & Nanotechnology, Bilkent University, Ankara, 06800, Turkey
    Authors for correspondence: *svempati01@qub.ac.uk; **uyar@unam.bilkent.edu.tr

ABSTRACT

A hierarchy of nanostructured-ZnO was fabricated on the electrospun nanofibers by atomic layer deposition (ALD) and hydrothermal growth, subsequently. Firstly, we produced poly(acrylonitrile) (PAN) nanofibers via electrospinning, then ALD process provided a highly uniform and conformal coating of polycrystalline ZnO with a precise control on the thickness (50 nm). In the last step, this ZnO coating depicting dominant oxygen vacancies and significant grain boundaries was used as a seed on which single crystalline ZnO nanoneedles (average diameter and length of ~25 nm and ~600 nm, respectively) with high optical quality were hydrothermally grown. The detailed morphological and structural studies were performed on the resulting nanofibers, and the photocatalytic activity (PCA) was tested with reference to the degradation of methylene blue. The results of PCA were discussed in conjunction with photoluminescence response. The nanoneedle structures supported the vectorial transport of photo-charge carriers, which is crucial for high catalytic activity. The enhanced PCA, structural stability and reusability of the PAN/ZnO nanoneedles indicated that this hierarchical structure is a potential candidate for waste water treatment.

INTRODUCTION

Development of novel materials with enhanced photocatalytic activity (PCA) along with stability is one of the intensely researched topics for water purification and waste treatment. The need for such research arises due to water pollution and ever increasing environmental issues threatening the human health severely [1-3]. Metal oxides such as ZnO in the structures of nanoparticles [4], nanorods [5, 6] and nanofibers [4] are widely studied for water purification purposes due to their well known PCA. On the other hand, electrospun polyacrylonitrile (PAN) nanofibers have been widely adopted in water filtration due to their unique properties including high surface area, nanoporous structure, low basis weight, easy permeability, good stability and chemical resistance [7-10]. Here we fabricated a hierarchy of nanostructured-ZnO depicting a synergy effect to enhance the PCA on electrospun PAN nanofibers using chemical vapor deposition and liquid phase deposition techniques, namely atomic layer deposition (ALD) and hydrothermal growth [11].

EXPERIMENTAL DETAILS

N,N-dimethylformamide (Pestanal, Riedel) was used as a solvent to prepare 12% (w/v) PAN (Mw: 150,000, Scientific Polymer Products, Inc.) solution. For the electrospinning of the PAN solution; feed rate, applied voltage and tip-to-collector distance were 1 mL/h, 15 kV and 12 cm, respectively. ZnO seed deposition on electrospun PAN nanofibers was carried out at ~200

°C in a Savannah S100 ALD reactor (Cambridge Nanotech Inc.) using diethylzinc (Sigma-Aldrich) and HPLC grade water as the zinc precursor and oxidant, respectively. N_2 was used as a carrier gas at a flow rate of ~20 sccm. 400 cycles were applied via exposure mode (a trademark of Ultratech/Cambridge Nanotech Inc.) in which dynamic vacuum was switched to static vacuum before each precursor pulse. For the growth of ZnO nanoneedles on 3.6 mg of PAN/ZnO seed by hydrothermal process (90 °C, 5 h) in crucible, equimolar (0.02 M) zinc acetate dihydrate (≥ 98%, Sigma-Aldrich) and hexamethylene tetramine (≥ 99%, Alfa Aesar) were used. The morphology of the samples was studied using a scanning electron microscope (SEM, FEI – Quanta 200 FEG) with a nominal 5 nm of Au/Pd sputter coating. Transmission electron microscope (TEM, FEI–Tecnai G2F30) images and selected area electron diffraction (SAED) pattern were also obtained. Photoluminescence (PL) measurements were performed using Horiba Scientific FL-1057 TCSPC at an excitation wavelength of 360 nm. Methylene blue (MB, Sigma-Aldrich, certified by the Biological Stain Commission) was used as a model organic dye to test PCA of the samples. The nanofibrous mats (weight: 3.6 mg) were immersed in quartz cuvettes containing the MB solution (18.8 µM). The cuvettes were exposed to UV light (300 W, Osram, Ultra-Vitalux, sunlight simulation) placed at a distance of ~15 cm. Dye concentrations in the cuvettes were measured using a UV-Vis-NIR spectrophotometer (Varian Cary 5000) at regular time intervals. We have repeated the PCA experiment twice (i.e. 2nd and 3rd cycles) for PAN/ZnO needle sample (~3.3 mg) to determine the reusability versus performance. All the figures are reproduced with permission from Ref [11]

DISCUSSION

ZnO seed-coated PAN nanofibers were fabricated through electrospinning and ALD processes, on which ZnO nanoneedles were hydrothermally grown [11]. Schematic representations of the electrospinning and ALD processes, and fabrication procedure for the hierarchical PAN/ZnO needle nanofiber are illustrated in figure 1.

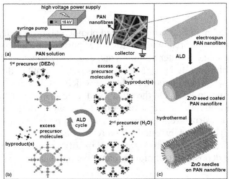

Figure 1. Schematic representations of the (a) electrospinning of PAN solution, (b) ALD of ZnO seed onto PAN nanofiber, and (c) fabrication procedure for hierarchical PAN/ZnO needle nanofiber.

The representative SEM images of PAN nanofibers are given in figure 2(a), where the average fiber diameter (AFD) is estimated to be ~655 ± 135 nm. After the ALD process, we have recorded the SEM images which are shown in figure 2(b) where the AFD is ~715 ± 125 nm. AFD increased because of ALD coating, and moreover the fiber structure was stable during the ALD process. Subsequently, the hydrothermal method was employed to grow ZnO nanoneedles on the ZnO seed-coated PAN nanofibers. Straight ZnO nanoneedles covered the surface of the ZnO seed-coated PAN nanofibers, and no branching was observed, figure 2(c1-2). By analyzing the figure 2(c2), average diameter and length of the nanoneedles were determined as ~25 nm and ~600 nm, respectively.

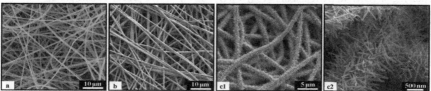

Figure 2. Representative SEM images of (a) pristine PAN, (b) PAN/ZnO seed, and (c1-2) PAN/ZnO needle nanofibers at different magnifications.

Despite the relatively large surface area of the nanofibers, the conformal coating of ZnO with a uniform thickness (~50 nm) was observed from the TEM image of PAN/ZnO seed nanofiber shown in figure 3(a1). For the high surface area substrates such as non-woven nanofiber mat, ALD is a well suitable technique as shown by us earlier [12, 13]. The SAED pattern in figure 3(a2) reveals the polycrystalline nature of ZnO seed. The bright spots on the polycrystalline diffraction rings were observed due to the presence of well crystalline grains [13]. Moreover, high resolution TEM (HRTEM) and fast Fourier transform (FFT) images indicate the single crystalline nature of ZnO nanoneedles (figure 3(b1), (b2)). It is important to determine the growth direction of ZnO; hence, the lattice spacing was measured as ~0.525 nm corresponding to c-axis that is the preferential growth direction of the nanoneedles. Therefore polar planes of ZnO have shown to depict relatively higher PCA [14].

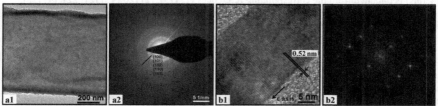

Figure 3. Representative (a1) TEM image and (a2) SAED pattern of PAN/ZnO seed nanofiber; (b1) HRTEM image and (b2) FFT image of ZnO needle.

We infered the information about surface defects from the PL spectra of nanofibers (figure 4a); such defects play a crucial role in determining the PCA of the material. Based on the literature, various crystal defects and the possible transitions with emission wavelengths are schematized in figure 4b. It is known that the typical excition emission band corresponding to the interband transition lies in the UV region for ZnO, while the defect related emission is in the visible region [16-18]. PAN/ZnO needle has shown a clear peak in UV region, while PAN/ZnO seed did not show any clear excition emission peak. Violet emissions were broader, and not as prominent as green emission, which are related to zinc interstitials (Zn_i) and oxygen vacancies (V_O), respectively. PAN/ZnO seed has the predominant visible emission because of the large area of grain boundaries. For PAN/ZnO needle, a slight blue shift can be noticed in the center of the peak in the visible region of the spectrum due to its seed counterpart. Therefore V_O's can create intermediate bands. Hence, the green emission is a combination of bulk grain region (BGR) and depletion region (DR). In BGR and in DR $V_O^+ \rightarrow V_O^*$ and $V_O^+ \rightarrow V_O^{++}$ processes take place, respectively. BGR of ZnO is not accessible to the PCA, since it is in deep inside the lattice. However depletion layer within grain boundaries is extremely helpful for PCA. The intensity ratio of UV to visible emission is ten times higher for the PAN/ZnO needle than PAN/ZnO seed. This high ratio indicates higher optical quality of the PAN/ZnO needle sample.

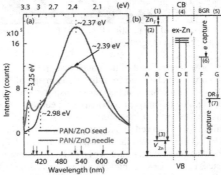

Figure 4. (a) PL spectra of PAN/ZnO seed and needle counterpart and (b) depicts various crystal defects and possible transitions [16]. The energetic location of each defect level (denoted by numerals) is obtained from the corresponding references (1) [19], (2) [20], (3) [21, 22], (4) [23], (5) [17, 18], (6) [24] and (7) [25]. The alphabets stand for emission wavelengths in nanometer, where A = 395, B = 437, C = 405, D = 440, E = 455, F = ~500, and G = 564. V_{Zn} is located 0.30 eV above the VB, while Zn_i is at 0.22 eV below the CB.

We have comparatively investigated the PCA of nanofibers (figure 5). Pristine MB solution data has shown a decay constant of ~157 min. On the other hand degradation rates clearly showed that the PCA was much higher for the needle sample (~14.6 min) compared to seed one (~113.1 min). Under a suitable illumination e and h pairs are created in VB. *Electrons* can be excited from the VB to reach the CB, leaving behind *holes* in the VB [16, 18]. before these separated charges recombine, they can migrate to the surface of the semiconductor, then they can participate in the redox reactions to form hydroxyl radical ($\cdot$OH) that is the key for the PCA [26-28]. Interband transition in the PL spectrum showed that PCA is taking place at CB and VB in defect free nanoneedle. In the case of seed sample, there was no clear interband transition in PL, hence the PCA taking place at CB and VB was not dominant. In contrast, DR is well accessible for PCA for this sample. For PAN/ZnO needle case, significantly higher PCA was yielded through the combination of all these processes. As a result, PAN/ZnO needle sample has shown higher PCA because of the higher surface area and higher crystal quality of the needle-morphology. Moreover catalytic activity occurs at surface defects on ALD seed and VB, and CB of nanoneedles. PCA experiment have been repeated twice for the PAN/ZnO needle, and the results showed almost no decay in the catalytic activity of this material when reused.

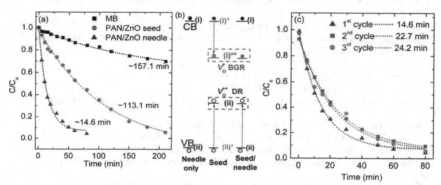

Figure 5. (a) Degradation rate of MB in aqueous environment tested for pristine, in the presence of PAN nanofibers, PAN/ZnO seed and PAN/ZnO needle (1st cycle) cases, (b) plausible mechanism of photocatalysis involving oxygen vacancies, where (i) and (ii) stand for processes acceptor → acceptor $^-$ and donor → donor $^+$ respectively, and (c) PCA of PAN/ZnO needle nanofibers for 1st, 2nd and 3rd cycles.

CONCLUSION

Here hydrothermal growth of single crystal, needle like structures of ZnO was performed on the ALD seed coated electrospun PAN nanofibers. We have reported the results of an investigation on ZnO based photocatalyst. The PCA improvement was attributed to the collective effect, which enabled the active participation of defect state and the catalysis taking place at CB, as well as VB. Moreover, the samples were subjected to recycling, and the PAN/ZnO needle depicted a comparable performance with the fresh sample. The performance, structural stability and reusability of these ZnO nanostructures make them potential candidates for waste water treatment with solar energy.

ACKNOWLEDGMENTS
S.V. thanks The Scientific & Technological Research Council of Turkey (TUBITAK) (TUBITAK-BIDEB 2216, Research Fellowship Programme for Foreign Citizens) for postdoctoral fellowship. F.K. thanks TUBITAK-BIDEB for a PhD scholarship. N.B. thanks EU FP7-Marie Curie-IRG for funding NEMSmart (PIRG05-GA-2009-249196). T.U. thanks EU FP7-Marie Curie-IRG (NANOWEB, PIRG06-GA-2009-256428) and The Turkish Academy of Sciences - Outstanding Young Scientists Award Program (TUBA-GEBIP) for funding. Authors thank M. Guler for technical support for TEM analysis.

REFERENCES

1. Q. Li, S. Mahendra, D. Y. Lyon, L. Brunet, M. V. Liga, D. Li and P. J. J. Alvarez, Water Res. 42, 4591 (2008).
2. F. Meng, S. R. Chae, A. Drews, M. Kraume, H. S. Shin and F. Yang, Water Res. 43, 1489 (2009).
3. M. M. Khin, A. S. Nair, V. J. Babu, R. Murugan and S. Ramakrishna, Energy Env. Sci. 5, 8075 (2012).
4. H. Liu, J. Yang, J. Liang, Y. Huang and C. Tang, J. Am. Ceram. Soc. 91, 1287 (2008).
5. A. Sugunan, V. K.Guduru, A. Uheida, M. S.Toprak and M. R. Muhammed, J. Am. Ceram. Soc. 93, 3740 (2010).
6. Z. Chang, Chem. Comm. 47, 4427 (2011).
7. N. Scharnagl, H. Buschatz, Desalination 139, 191 (2001).
8. S. Yang, Z. Liu, J. Membr. Sci. 222, 87 (2003).
9. L. Zhang, J. Luo, T.J. Menkhaus, H. Varadaraju, Y. Sun, H. Fong, J. Membr. Sci. 369, 499 (2011).
10. Y. Mei, C. Yao, K. Fan, X. Li, J. Membr. Sci. 417, 20 (2012).
11. F. Kayaci, S. Vempati, C. Ozgit-Akgun, N. Biyikli, T. Uyar, Applied Catalysis B: Environmental, http://dx.doi:10.1016/j.apcatb.2014.03.004, (2014).
12. F. Kayaci, C. Ozgit-Akgun, I. Donmez, N. Biyikli, T. Uyar, ACS Appl. Mater. Interfaces, 4, 6185 (2012).
13. F. Kayaci, C. Ozgit-Akgun, N. Biyikli, T. Uyar, RSC Adv. 3, 6817 (2012).
14. J. Wang, P. Liu, X. Fu, Z. Li, W. Han, X. Wang, Langmuir, 25 1218 (2009).
15. S. Cho, J.-W. Jang, J.S. Lee, K.-H. Lee, Nanoscale, 4, 2066 (2012).
16. S. Vempati, J. Mitra, P. Dawson, Nanoscale Res. Lett. 7, 470 (2012).
17. J.D. Ye, S.L. Gu, F. Qin, S.M. Zhu, S.M. Liu, X. Zhou, W. Liu, L.Q. Hu, R. Zhang, Y. Shi, Y.D. Zheng, Appl. Phys. A: Mater. Sci. Process., 81, 759 (2005).
18. S. Vempati, S. Chirakkara, J. Mitra, P. Dawson, K.K. Nanda, S.B. Krupanidhi, Appl. Phys. Lett. 100, 162104 (2012).
19. C.H. Ahn, Y.Y. Kim, D.C. Kim, S.K. Mohanta, H.K. Cho J. Appl. Phys. 105, 013502 (2009).
20. E.G. Bylander, J. Appl. Phys. 49, 1188 (1978).
21. B. Lin, Z. Fu, Y. Jia, Appl. Phys. Lett. 79, 943 (2001).
22. P.S. Xu, Y.M. Sun, C.S. Shi, F.Q. Xu, H.B. Pan, Nucl. Instrum. Methods B 199 286 (2003).
23. H. Zeng, G. Duan, Y. Li, S. Yang, X. Xu, W. Cai, Adv. Funct. Mater. 20, 561 (2010).
24. K. Vanheusden, W.L. Warren, C.H. Seager, D.R. Tallant, J.A. Voigt, B.E. Gnade, J. Appl. Phys. 79, 7983 (1996).
25. A.v. Dijken, E.A. Meulenkamp, D. Vanmaekelbergh, A. Meijerink, J. Lumin. 90, 123 (2000).
26. N. Daneshvar, D. Salari, A.R. Khataee, J. Photochem. Photobiol. A 162, 317 (2004).
27. R.W. Matthews, J. Catal. 97, 565 (1986).
28. Izumi, W.W. Dunn, K.O. Wilbourn, F.R.F. Fan, A.J. Bard, J. Phys. Chem. 84, 3207 (1980).

Mater. Res. Soc. Symp. Proc. Vol. 1675 © 2014 Materials Research Society
DOI: 10.1557/opl.2014.847

Electrosynthesized Polystyrene Sulphonate-Capped Zinc Oxide Nanoparticles as Electrode Modifiers for Sensing Devices

Maria C. Sportelli[1], Diana Hötger[2], Rosaria A. Picca[1], Kyriaki Manoli[1], Christine Kranz[2], Boris Mizaikoff[2], Luisa Torsi[1], and Nicola Cioffi[1]

[1]Dipartimento di Chimica, Università degli Studi di Bari "Aldo Moro", Via E. Orabona 4, 70126 Bari, Italy.
[2]Institut für Analytische und Bioanalytische Chemie, Universität Ulm, Albert-Einstein-Allee 11, 89081 Ulm, Germany.

ABSTRACT

ZnO nanoparticles were prepared by a green electrochemical synthesis method applying low current densities followed by a thermal treatment. Sodium polystyrene sulphonate (PSS) was used as stabilizer in the electrolytic aqueous medium due to its biocompatibility and stability. The as-prepared nanocolloids were then annealed to improve their stability, and then converted via hydroxide species into stoichiometric oxide. Different calcination temperatures were studied. ZnO@PSS nanomaterials were deposited onto SiO_2/Si substrates, in part in combination with an organic semiconductor layer to evaluate their influence on organic field effect transistors (OFETs). All nanomaterials and composite layers were characterized by morphological and spectroscopic techniques. Promising results regarding the use of ZnO@PSS in OFETs could be demonstrated.

INTRODUCTION

Zinc oxide nanoparticles (ZnO-NPs) may offer excellent prospects for designing a new generation of low-cost, flexible, multi-functional bioelectronic devices. The conjugation of the high surface area-to-volume ratio of ZnO-NPs with their affinity for biomolecules, electron communication ability, and chemical stability already rendered them a particularly promising material for biosensing applications [1,2].

Typical approaches for the synthesis of ZnO-NPs involve sol-gel [3] and hydrothermal methods [4], while electrochemical routes are less explored, although they may offer several advantages for fine-tuning of particle morphology, size, surface chemistry, etc. [5]. We have recently developed an electrochemical preparation routing of ZnO-NPs from aqueous media in the presence of several stabilizers including sodium polystyrene sulphonate (PSS). The thermal treatment of the nanocolloid at $t \geq 300°C$ allows the complete conversion into nanostructures having a ZnO stoichiometry.

PSS was preferred to other anionic capping agents considering its widespread use in organic field effect transistors (OFETs) [6] and its biocompatibility. Moreover, its sulphonated moiety can be used for further functionalizing the ZnO@PSS nanophases via ionic interactions with cationic (bio)molecules. In this work, the as-prepared nanostructures were employed as electrode modifiers in OFETs showing improved electrical performances. A systematic analytical characterization by transmission electron microscopy (TEM), UV-Vis, IR, and X-ray photoelectron (XPS) spectroscopies was performed to investigate the physicochemical properties of ZnO@PSS NPs and the resulting composite layers.

EXPERIMENT

Materials

Zn sheets (purity 99.99+%) and poly(3-hexylthiophene-2,5-diyl) (P3HT, purity 99.995%) were from Goodfellow Ltd and Rieke Metals Inc., respectively. All the other chemicals were of analytical grade and purchased from Sigma-Aldrich.

Synthesis of ZnO-NPs

ZnO nanocolloids were electrosynthesized in an alkaline solution of $NaHCO_3$ (30 mM) [7] in the presence of PSS ($M_w \sim 70000$, 1 g/L). The electrosynthesis was carried out galvanostatically at 10 mA/cm^2 for 1 h using a CH-1140b potentiostat-galvanostat (CH Instruments Inc., Texas) in a three-electrode cell configuration with two zinc plates as working and counter electrodes, respectively, and Ag/AgCl (KCl sat.) as reference electrode. After centrifugation (at 5000 rpm for 30 min), the resulting precipitate was dried at 70°C, overnight. The obtained powder was calcined either at 300°C or at 600°C in a muffle furnace in air for 1 h.

Analytical characterization

As-prepared and calcined samples were morphologically characterized by TEM on a FEI Tecnai 12 system operating at HV 120 kV. XPS surface analysis was performed on a Theta Probe spectrometer (Thermo Fisher Scientific), equipped with a monochromatized AlKα source, beam spot size = 300 μm. UV-Vis and Fourier Transform Infrared (FTIR) characterizations were carried out on a Shimadzu UV-1601 double beam spectrometer and on a BioRad FTS6000 instrument, respectively.

OFET preparation and testing

OFET devices were prepared on thermally grown 300 nm SiO_2/p-doped Si substrate (gate electrode, G). A mixed solution consisting in 600°C-calcined ZnO NPs (1 g/L) and P3HT (2.5 g/L) organic semiconductor was deposited by spin coating (2000 rpm, for 1 minute) forming a mixed organic/inorganic semiconducting layer. The electrical characteristics of the fabricated devices were tested using an Agilent 4155C semiconductor parameter analyzer.

DISCUSSION

In this work, the electrochemical method firstly reported in [7] has been modified by adding a stabilizing agent to the electrolytic medium, which resulted in an improved morphology and control of NP growth. Cationic (cetyl-trimethylammonium bromide, CTAB), neutral (polyvinylpyrrolidone, PVP), and anionic (PSS) stabilizers were selected to explore their effect on the yield of the process (Table I).

Table I. Process efficiency for the preparation of nanocolloids. Expressed in terms of nanodispersed Zn mass. This value is calculated by mass variation of anode (Δm_a) and cathode (Δm_c), weighting the electrodes before and after the electrosynthesis. Cell volume = 7 mL.

Stabilizer in NaHCO$_3$ 30 mM	Colloid concentration[*] [g/L]
None	1.63 ± 0.12
CTAB 0.1 M	2.57 ± 0.08
PVP 0.005 M	3.10 ± 0.07
PSS 1 g/L	3.18 ± 0.11

It was evident that all tested stabilizers significantly improved the reaction yield with PSS showing the best performance. Moreover, a pH increase was always observed. The formation of various Zn(II) hydroxides and mixed hydroxyl-carbonate species was detected by the milky color observed during the electrosynthesis. A thermal treatment was then introduced to achieve the complete conversion of the produced colloid into ZnO-NPs. The influence of capping agent and annealing temperature on NP morphology was studied by TEM characterization.

TEM analysis

The effects of different stabilizers on the ZnO nanomaterials' morphology are shown in fig. 1. In fact, 300°C-calcined ZnO samples prepared in the absence (a) or in the presence of CTAB (b) or PVP (c) resulted in different morphologies (from amorphous to rod-like and platelet-like structures), though the degree of morphological control remained limited. In contrast, PSS appears to be a suitable candidate allowing ZnO-NP morphological control both in the pristine sample (d), and after calcination at 300°C (e) and 600°C (f). Evidently, the average NP diameter of ZnO-NPs increased only slightly from pristine to annealed samples. As a result, PSS was preferred for further studies and applications of ZnO-NPs.

Figure 1. TEM images of electrosynthesized ZnO-based materials: a) no stabilizer, calcined at 300°C; b) CTAB as stabilizer, calcined at 300°C; c) PVP as stabilizer, calcined at 300°C; d) PSS

as stabilizer, pristine; e) PSS as stabilizer, calcined at 300°C; f) PSS as stabilizer, calcined at 600°C.

UV-Vis, FTIR, XPS spectroscopic characterizations

The slight increase of PSS-stabilized NP diameter after calcination was also confirmed by UV-Vis experiments (Fig. 2a) showing that the absorption maximum typical of nanostructured ZnO was shifted from 355 nm for the freshly prepared nanocolloids (dashed line) to 376 nm for 600°C-calcined ZnO-NPs (solid line). Nonetheless, this shift was lower than that reported for calcined particles prepared without stabilizer [7]. Regarding the chemical composition of pristine and calcined samples, FTIR and XPS spectroscopy were particularly useful for the identification of the different chemical environments associated to the electrochemical and calcination processes. IR analysis (Fig. 2b) showed strong bands associated to the presence of carbonate (1381 cm^{-1}), adsorbed water and/or hydroxyl species (O-H stretching at 3385 cm^{-1}, bending at 1502 cm^{-1}), and a small signal around 460 cm^{-1} (Zn-O stretching) in the pristine samples. In particular, the bands at 1381 and 1502 cm^{-1} are typically associated to hydrozincite-like species $(Zn_x(CO_3)_y(OH)_z)$ [8]. Furthermore, the increase in the calcination temperature led to the gradual disappearance of carbonate and hydroxyl signals, and the parallel increase of the signal relevant to Zn-O stretching, thus confirming the conversion to ZnO.

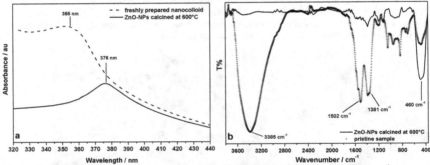

Figure 2. a) UV-Vis and b) IR spectra of pristine and 600°C-calcined ZnO-NPs samples.

Interestingly, a similar behavior was found at the sample surface considering XPS characterization. XPS data showed that the Zn2p$_{3/2}$ XP region (Fig. 3) could be deconvoluted into two components for pristine samples (attributable to hydroxyl and carbonate/oxide species [9]) and only one signal, compatible with the presence of ZnO that appeared after calcination (t ≥ 300°C). Thus, obtained data were in good agreement with XPS results obtained at ZnO standard powder, and curve fitting of the ZnL$_3$M$_{45}$M$_{45}$ Auger signal.

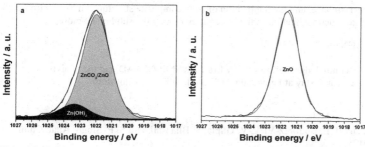

Figure 3. Zn2p$_{3/2}$ XP spectra of a) pristine, and b) 600°C-calcined ZnO-NPs samples.

Application of ZnO-NPs in OFETs

Preliminary experiments regarding the combination of ZnO-NPs with OFETs were carried out by mixing P3HT - an organic semiconductor typically used in our laboratories for bio-OFET development [10] - with calcined ZnO-NPs in chloroform. The OFET channel length (L) was 200 μm, while the channel width (W) was 4 mm. The electrical performance of the OFETs was evaluated by measuring current–voltage (I-V) characteristic curves (I$_{DS}$ versus V$_{DS}$ at different gate bias). The values of mobility (μ), on-off current ratio (I$_{ON}$/I$_{OFF}$), and threshold voltage (V$_T$) ,extracted from the I-V characteristics [11], are reported in Table II. Both P3HT layer and ZnO-NPs/P3HT mixed layer were tested for comparison purposes. It was shown that the addition of ZnO-NPs resulted in an enhancing effect in FET mobility. However, undesirable side effects were also evident, such as the decrease in I$_{ON/OFF}$ and the increase of V$_T$ value. The devices were very robust, as even after weeks a high mobility was still observed.

Table II. Average values of mobility, I$_{OFF}$/I$_{OFF}$ ratio and the threshold voltage obtained from devices fabricated with bare P3HT, and ZnO-NPs/P3HT.

Sample	Mobility μ [cm^2/Vs]	V$_T$ [V]	I$_{ON/OFF}$
Bare P3HT layer	$(4.4 \pm 0.7) \, 10^{-4}$	29 ± 5	27 ± 18
Mixed ZnO/P3HT	$(2.3 \pm 0.4) \, 10^{-3}$	38 ± 12	17 ± 9

CONCLUSIONS

A novel, facile, and green electrochemical approach was developed for the preparation of ZnO-NPs. Different stabilizers were employed in the electrosynthesis step proving PSS to be the more effective in terms of reaction yield and NP morphology. A calcination step at t ≥ 300°C was necessary to convert hydroxides/carbonates mixed species formed during the synthesis into pure ZnO-NPs. This step led to a slight increase in NP size. TEM, UV-Vis, FTIR, and XPS characterizations confirmed such results, showing that 600°C-calcined NPs were suitable candidates for application in OFETs. Preliminary experiments performed on devices prepared by spin coating deposition of a mixed layer made of ZnO-NPs/P3HT showed promising results in terms of higher mobility and device stability, though I$_{ON}$/I$_{OFF}$ ratio and V$_T$ need improvements. Future work is targeted towards further investigation of the possible enhancement effect of ZnO-

NPs on the sensor performance, and the application of such composite structures in bioelectronic devices via appropriate functionalization of the material exploiting the sulphonate moiety.

ACKNOWLEDGMENTS

Diana Hötger acknowledges fellowship by Erasmus SMP / DAAD and the analytical group of GDCh for a research stay at Università degli Studi di Bari "Aldo Moro".

REFERENCES

1. S. K. Arya, S. Saha, J. E. Ramirez-Vick, V. Gupta, S. Bhansali, and S. P. Singh, Anal. Chim. Acta **737**, 1 (2012).
2. M. Tak, V. Gupta, and M. Tomar, Biosens. Bioelectron. **59**, 200 (2014).
3. E. A. Meulenkamp, J. Phys. Chem. B **102**, 5566 (1998).
4. S. Baruah and J. Dutta, Sci. Technol. Adv. Mater. **10**, 013001 (2009).
5. M.C. Sportelli, S. Scarabino, R.A. Picca, and N. Cioffi, in *CRC Concise Encycl. Nanotechnol.* (Taylor & Francis Group), submitted.
6. P. Lin and F. Yan, Adv. Mater. **24**, 34 (2012).
7. K. G. Chandrappa, T. V. Venkatesha, K. Vathsala, and C. Shivakumara, J. Nanoparticle Res. **12**, 2667 (2010).
8. M. C. Hales and R. L. Frost, Polyhedron **26**, 4955 (2007).
9. National Institute of Standards and Technology, NIST X-Ray Photoelectron Spectroscopy Database Version 4.1 (2012).
10. M. D. Angione, S. Cotrone, M. Magliulo, A. Mallardi, D. Altamura, C. Giannini, N. Cioffi, L. Sabbatini, E. Fratini, P. Baglioni, G. Scamarcio, G. Palazzo, and L. Torsi, Proc. Natl. Acad. Sci. **109**, 6429 (2012).
11. L. Torsi and A. Dodabalapur, Anal. Chem. **77**, 380 A (2005).

Mater. Res. Soc. Symp. Proc. Vol. 1675 © 2014 Materials Research Society
DOI: 10.1557/opl.2014.861

Effects of Annealing on Structural and Optical Properties of ZnO Nanowires

Anas Mazady, Abdiel Rivera, and Mehdi Anwar

*Electrical and Computer Engineering, University of Connecticut, Storrs, CT 06269, Email:
anwara@engr.uconn.edu*

ABSTRACT

We report, for the first time, effects of annealing of ZnO NWs grown on p-Si substrates. ZnO NWs are grown using metalorganic chemical vapor deposition (MOCVD) and thermal annealing was performed *in situ* under nitrogen ambient at different stages of the growth process. Increasing the annealing temperature of the ZnO seed epi-layer from 635 °C to 800 °C does not affect the morphology of the grown NWs. In contrast, annealing the NWs themselves at 800 °C results in a 48% decrease of the surface area to volume ratio of the grown NWs. The optical quality can be improved by annealing the seed layer at a higher temperature of 800 °C, although annealing the NWs themselves does not affect the defect density.

INTRODUCTION

ZnO is a promising material for optoelectronic applications, such as laser diodes, optical sensors, solar cells, light emitting diodes (LEDs), and quantum cascade lasers (QCLs) due to its large direct band gap energy of 3.37 eV and a large excitonic binding energy of 60 meV [1, 2]. ZnO nanowires (NWs) exhibit large spontaneous and strain-induced piezoelectric polarizations required for energy harvesting applications [3]. In addition, being chemically stable and biocompatible, ZnO NWs-based devices can be used for bio-sensing applications [4]. Growth of ZnO nanowires (NWs) has been reported on different substrates, with Si substrates being of particular interest in order to maintain CMOS process compatibility [5]. However, the large lattice mismatch of 40% and large difference in the thermal expansion coefficients of 87% between ZnO and Si substrates cause built-in residual stress in the grown ZnO NWs. Proper annealing conditions, as has been investigated in this paper, can reduce this built-in residual stress and hence can improve the crystalline quality and minimize defects of the grown ZnO NWs.

Native defects such as Zn and O vacancies vary upon growth conditions depending on the formation energy. These defects can degrade the crystal and optical quality of the (film or NWs). Nevertheless, these defects can be suppressed by post annealing process depending on the temperature and atmosphere (under air, N, argon, etc). Borseth et al. [6] demonstrated suppression of native defect, such as, V_{Zn}^- and V_O^+, by annealing under O-rich, Zn-rich, and under a mixture of both atmospheres. Chen et al. [7] used positron annihilation spectroscopy to study vacancy defects in ZnO as a function of annealing temperature. The annealing experiment showed that the V_{Zn}^- defects suppress at 600 °C while annealing at temperatures above 1000 °C produces these defects again. Considering the short windows to passivate native defects on ZnO, we investigated the impact of annealing on ZnO NWs before and after the growth in order to improve the crystal and optical quality.

EXPERIMENT

ZnO NWs are grown on p-Si substrates using First Nano EasyTube 3000 MOCVD system. The NWs growth was preceded by a ZnO epilayer growth for 20 min at a constant temperature and pressure of 300 °C and 70 Torr, respectively, to achieve catalyst free growth and better crystalline quality. Diethylzinc (DEZn) and N_2O were used as the Zn and oxygen precursors with flow rates of 50 SCCM and 35 SCCM, respectively. For the NWs growth step DEZn and N_2O flow rates of 20 SCCM and 285 SCCM were maintained for 1 hr. N_2 was used as the career gas in all the processing steps. Three different annealing conditions were investigated. The epitaxial layer was annealed at 635 °C (ETA635) and 800 °C (ETA800) in the first two sets of samples without a post-annealing treatment after the NWs growth. The third set of samples was annealed at 800 °C both after the epilayer and NWs growth (ENTA800). Annealing of all the samples was performed *in situ* under nitrogen ambient.

DISCUSSION

Scanning electron microscope (SEM) images of the NWs annealed under different condition are shown in Figure 1. It can be observed that while annealing the epilayer at different temperatures (ETA635 and ETA800) does not affect the NWs morphology, annealing the NWs at 800 °C (ENTA800) deteriorates the NWs morphology by decreasing the surface area to volume ratio. The surface area to volume ratio is 0.082 nm^{-1} for both ETA635 and ETA800 with average lengths and diameters of 1 μm and 50 nm, respectively. For the annealed NWs, this ratio decreases by 48% to 0.043 nm^{-1}, owing to a modification of the average length and diameter to 700 nm and 100 nm, respectively. Annealing the epitaxy does not affect the ZnO NWs morphology because the NWs are already lattice matched with the ZnO epitaxial layer. The decrease in surface area to volume ratio for annealing the NWs may be related to the decrease in c-lattice constant of the NWs which is discussed in the later part of the paper. Electron dispersive X-ray spectroscopy (EDS) suggested growth of near-stoichiometric ZnO with elemental composition of 44% oxygen and 56% zinc.

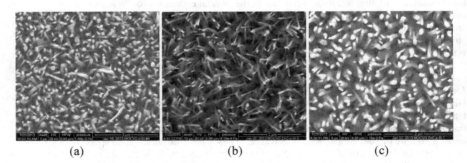

(a)   (b)   (c)

Figure 1. SEM images of ZnO NWs with different annealing condition: (a) ETA635, (b) ETA800, (c) ENTA800

X-ray diffraction (XRD) measurements of the samples were performed using a Brunker D8 Advanced equipment with CuK$_\alpha$ excitation at λ=1.5406 Å. XRD data for ETA635 and ENTA800 are compared in Figure 2. A dominant peak along (002) orientation and very small/ no peaks along the other growth directions suggest single crystalline growth of ZnO NWs in all the cases. The inset of Figure 2 shows a magnified view of only the (002) peaks. Lorentzian fitting algorithm was performed to decompose the (002) peaks. For the ETA635 sample the peak at 34.414° is due to the NWs and the peak at 34.415° is due to the ZnO seed layer. For the ENTA800 sample the peaks at 34.443° and 34.5° represent ZnO NWs and thin film, respectively. The following relationship was used to calculate c-lattice constants of the ZnO NWs:

$$\sin^2 \theta = \frac{\lambda^2}{4}\left[\frac{4}{3}\left(\frac{h^2 + hk + k^2}{a^2}\right) + \frac{l^2}{c^2}\right]$$

Where, θ is the angle of diffraction, *(h, k, l)* are the lattice indices, and *a, c* are the lattice constants. The c-lattice constants for the ETA635 and ENTA800 samples were calculated to be 5.2078Å and 5.2036Å, respectively, while that for unstrained ZnO is 5.206Å [8]. The larger value of the c-lattice constant for the ETA635 sample suggests that the unit cell is elongated along the c-axis and a compressive strain is induced along the film plane (horizontal) which is in agreement with Lee et al. [9]. The decrease in c-lattice constant caused by a high temperature annealing is also in accordance with the observation made by Lee et al [9].

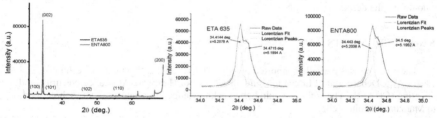

Figure 2. XRD measurement of ETA635 and ENTA800 samples. Split of the (002) peak due to the presence of both ZnO thin films (seed layer) and NWs are shown in the two insets.

Room temperature photoluminescence (PL) measurement was carried out to characterize the defect/ trap densities, as shown in Figure 3. A He-Cd laser line at 325 nm was used to excite the samples and an optical chopper, iHR 550 monochromator, and a SR830 lock-in amplifier were used in standard configuration to measure PL. The peaks at 650 nm and 975 nm are due to the second and third harmonics of the incident laser. The peak at ~388 nm represents near band-edge emission due to recombination of free excitons. The remaining peaks are due to oxygen vacancy and zinc vacancy related defects. As observed from the figure, annealing the NWs does not affect the trap counts (ETA800 and ENTA800 have similar PL spectra). In contrast, annealing the epitaxy at higher temperatures results in fewer traps in the NWs.

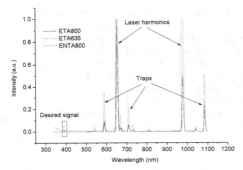

Figure 3. Room temperature PL spectra of ZnO NWs treated under different annealing conditions

CONCLUSIONS

ZnO NWs are grown using MOCVD with various post-growth annealing conditions. High temperature annealing of the ZnO seed layer does not affect the morphology of the grown NWs. Annealing the NWs results in a lower surface area to volume ratio. The optical quality gets better by annealing the seed layer at high temperatures, although annealing the NWs themselves does not affect the defect density.

ACKNOWLEDGMENTS

The research was partially supported by NSF I/UCRC, Magnolia Optical Technologies, Inc., NAVAIR, Department of Education, and the Center for Hardware Assurance, Security, and Engineering (CHASE). The authors would like to acknowledge the support of Drs. Ashok Sood, Tariq Manzur and John Zeller.

REFERENCES

[1] A. Rivera, J. Zeller, A. Sood and M. Anwar, "A Comparison of ZnO Nanowires and Nanorods Grown Using MOCVD and Hydrothermal Processes," *J Electron Mater,* vol. 42, pp. 894-900, 2013.

[2] H. C. Chou, A. Mazady, J. Zeller, T. Manzur and M. Anwar, "Room-Temperature Quantum Cascade Laser: ZnO/Zn$_{1-x}$Mg$_x$O Versus GaN/Al$_x$Ga$_{1-x}$N," *J Electron Mater,* vol. 42, pp. 882-888, 2013.

[3] J. Song, J. Zhou and Z. L. Wang, "Piezoelectric and semiconducting coupled power generating process of a single ZnO belt/wire. A technology for harvesting electricity from the environment," *Nano Letters,* vol. 6, pp. 1656-1662, 2006.

[4] J. Wang, X. W. Sun, A. Wei, Y. Lei, X. Cai, C. M. Li and Z. L. Dong, "Zinc oxide nanocomb biosensor for glucose detection," *Appl. Phys. Lett.,* vol. 88, pp. 233106-233106-3, 2006.

[5] R. Cross, M. De Souza and E. S. Narayanan, "A low temperature combination method for the production of ZnO nanowires," *Nanotechnology,* vol. 16, pp. 2188, 2005.

[6] T. M. Børseth, B. Svensson, A. Y. Kuznetsov, P. Klason, Q. Zhao and M. Willander, "Identification of oxygen and zinc vacancy optical signals in ZnO," *Appl. Phys. Lett.,* vol. 89, pp. 262112, 2006.

[7] Y. Chen, M. Cao, T. Wang and Q. Wan, "Microwave absorption properties of the ZnO nanowire-polyester composites," *Appl. Phys. Lett.,* vol. 84, pp. 3367-3369, 2004.

[8] Ü. Özgür, Y. I. Alivov, C. Liu, A. Teke, M. Reshchikov, S. Doğan, V. Avrutin, S. Cho and H. Morkoc, "A comprehensive review of ZnO materials and devices," *J. Appl. Phys.,* vol. 98, pp. 041301, 2005.

[9] Y. Lee, S. Hu, W. Water, K. Tiong, Z. Feng, Y. Chen, J. Huang, J. Lee, C. Huang and J. Shen, "Rapid thermal annealing effects on the structural and optical properties of ZnO films deposited on Si substrates," *J Lumin,* vol. 129, pp. 148-152, 2009.

Mater. Res. Soc. Symp. Proc. Vol. 1675 © 2014 Materials Research Society
DOI: 10.1557/opl.2014.830

Electrodeposited Cu2O|ZnO Heterostructures With High Built-In Voltages For Photovoltaic Applications

Shane Heffernan[1] and Andrew J. Flewitt[1]

1Electrical Engineering Division, Department of Engineering, University of Cambridge, JJ Thomson Avenue, Cambridge CB3 0HA

ABSTRACT

Methods of improving low-cost $Cu_2O|ZnO$ heterojunction diodes fabricated through galvanostatic deposition of Cu_2O are presented. Improved processing parameters responsible for maximizing built-in voltage (V_{bi}) are determined. The relationship between pH, deposition current, temperature, and diode quality is analyzed and a process window for optimal Cu_2O deposition on ZnO is obtained with a pH range between 12.0 and 12.1 and a current density range which is determined by the effect of both pH and deposition current (J_{dep}) on grain size. The pH window is found to be narrower than previously reported[1] and much narrower than the processing window for the deposition of Cu_2O films. A two-step approach deposition based on the use of different J_{dep} is presented for the first time. A V_{bi} of 0.6 V is achieved, which is the highest reported for cells produced using low temperature processing routes involving electrodeposition and reactive sputtering.

INTRODUCTION

The growth of small to medium scale off-grid power systems requires power solutions comprising Earth-abundant, non-toxic, and air-stable materials. Cuprous oxide (Cu_2O) thin film photovoltaic (PV) devices are compatible with such deployment due to the ultra-low raw material cost and potential for ambient processing[2–4]. Cu_2O has the potential to reach 20% power conversion efficiency based on the Shockley-Queisser limit[5,6] though homojunction and Schottky-barrier cells have long been precluded due to Cu_2O's lack of susceptibility to doping as well as a tendency for Cu migration to occur across the metal-semiconductor interface in Schottky-barrier cells[7]. Cell architectures have therefore thus far been limited to oxide heterojunctions such as $Cu_2O|ZnO$, for which the best thin-film cell efficiencies reported are in the 1-3% range[8,9]. In all such reported cases, the Cu_2O system is severely limited by poor fill factors and an open-circuit voltage (V_{oc}) which is far below the theoretical limit. This is largely due to a low built-in potential (V_{bi}) caused by non-ideal band alignment between the absorbing Cu_2O and ZnO window layer, and a high recombination-current driven by a large interface-trap capture cross-section[10,11]. In this work, evidence of a refined processing window for the electrodeposition of Cu_2O on ZnO is presented. Electrodeposition is the highest performing of the few cost-effective and demonstrably scalable techniques for fabricating Cu_2O solar cells capable of widespread deployment[8,9,12]. Both the pH of the electrolyte bath and the deposition current under galvanostatic control have established windows within which device performance has been optimized[9,13–15] though little thorough investigation into their role in diode quality has been reported. Understanding the effect of processing parameters on diode formation allows the interface to be optimized and subsequently allows more sophisticated deposition profiles, such as the two-step deposition discussed herein.

EXPERIMENT

Cu₂O diodes and solar cells were fabricated according to the architecture shown in Figure 1. Diodes are fabricated on 70-100 Ω $\square^{-1}$ ITO-coated glass (Sigma Aldrich) with pixel areas ranging from 0.2 to 0.8 mm². Cells were fabricated on <10 Ω $\square^{-1}$ ITO-coated glass (Praezision Optik) with pixel areas fixed at 12mm². 80nm ZnO:Al was sputtered using an RF magnetron sputterer from a ceramic target (ZnO; 1 wt.% Al) using a 35 sccm flow of Ar at an operating pressure of 7.4 ×10⁻³ mbar. 50 nm of undoped ZnO was sputtered using a high target utilisation sputtering (HiTUS) system, details of which have been previously reported[16].

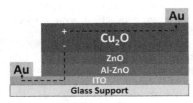

Figure 1: Solar cell device architecture

Cu₂O was deposited in a bath consisting solely of ~0.3 M CuSO₄ and 1.3 M Lactic Acid with NaOH added from 4.5 M stock to deliver the required pH. A Pt mesh and a Saturated Calomel Electrode (SCE) were used as counter and reference electrodes respectively. All depositions were carried out under galvanostatic control at 40 °C. All chemicals used were reagent grade. Gold contacts were evaporated thermally.

Diodes were tested using an Agilent B1500A SD Analyser. Current-voltage data shown was retrieved using 200 ms delay time with samples taken at 1 s and 5 s to confirm accuracy. Capacitance measurements were taken using a voltage perturbation frequency of 1 kHz. Solar cells were tested using an AM 1.5 filter with spectral mismatch retrieved from External Quantum Efficiency measurements and accounted for. Images were taken on a Philips Leo 1530 VP Scanning Electron Microscope.

RESULTS & DISCUSSION

Current density and pH in galvanostatic deposition

Electrodeposition of Cu₂O is achieved from an alkaline copper sulfate bath where the prevalence of OH⁻ ions encourages the formation of oxides from the Cu²⁺ cation present in the bath[17]. The equilibria between all possible species in such a bath are well documented[18]; the complexed Cu²⁺ ions present may undergo several redox processes that result in a solid precipitated species, either on the electrode substrate or in solution[1,18]. Previous work has established a wide pH window for deposition of single-phase, compact thin films of Cu₂O between approximately 9.0 and 12.5, with device performance and grain size purportedly maximized at the upper pH limit of 12.5 while further alkalization leads to Cu(OH)₂ formation[12,13,19]. The current or potential employed for deposition is also known to affect grain size and device performance. Further elucidation of the exact role that both bath parameters play in affecting diode quality requires a more rigorous approach of testing the effects of electrical bias at multiple values of pH. Figure 2 shows the effect that both pH and the deposition current (J_{dep}) have on the ratio of forward current to reverse current (on/off ratio) on diodes fabricated through galvanostatic electrodeposition of Cu_2O on compact thin films of undoped ZnO.

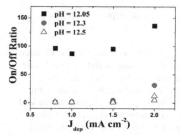 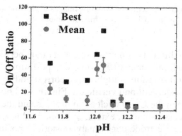

Figure 2a: Maximum On/off ratio for $Cu_2O|ZnO$ diodes at ±0.5 V bias for varying pH and J_{dep}.

Figure 2b: On/off ratios for $Cu_2O|ZnO$ diodes for varying pH. Cu_2O is deposited at 1 mA cm^{-2}.

Diode behavior is exhibited over a pH range from 11.6 to 12.2, with more alkaline solutions resulting in a loss of rectification in the diode and, in most cases, short circuiting of the Cu_2O film through pinholes resulting from the large grain size experienced at higher pH. Optimal diode behavior occurs in a relatively narrow range from 12.0 to 12.1 with both mean and maximum performance of devices exhibiting superior performance in this range. Figure 3 shows that the fall off in performance originates from higher series resistances in forward bias in the case of lower pH diodes, while diodes suffer from higher saturation currents at a higher pH.

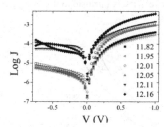

Figure 3: LogJ-V plots for diodes under varying pH.

These data suggest that a lower pH delivers better diode interfaces but more resistive films, while higher pH solutions suffer from poorer interfaces. SEM images in Figure 4 also verify the variation in grain sizes at different pH. It is likely that a lower pH results in a more favorable nucleation regime in which defect density is lower but the resulting grain density is larger while a higher pH increases the free energy of formation for Cu_2O nuclei, giving larger grains at the expense of a higher defect concentration at the interface.

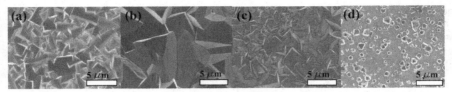

Figure 4: Scanning Electron Microscope Images of electrodeposited Cu_2O films. (a) -1 mA cm^{-2} at pH of 12.05; (b) -1 mA cm^{-2} at pH of 12.5; (c) -2 mA cm^{-2} at pH of 12.5; (d) -1 mA cm^{-2} at pH of 12.05 with deposition stopped during the occurrence of ΔE_{cell}.

The relationship between grain size and diode performance is further evidenced by the distinct, albeit limited, influence of J_{dep} on diode performance at fixed pH. Figure 2a shows a moderate increase in on/off ratio at higher J_{dep}, corresponding to a moderate decrease in grain size, as previously observed[1] and verified in Figure 4c. The results suggest that a critical relationship exists between the Cu_2O nucleation process and the resulting diode quality. The nucleation and growth process can, to an extent, be monitored *in situ* via the cell potential (E_{cell}) experienced by the reference electrode during galvanostatic deposition. Figure 5a shows a direct correlation between a characteristic shift to lower voltage (ΔE_{cell}) under constant J_{dep} after a certain amount of charge, typically 1-2 Coulombs, has passed through the cell. Notably, the magnitude of ΔE_{cell} scales consistently with the resulting photovoltaic performance of the device. The shift is found to mark the onset of growth over nucleation or the coalescence of grains as shown in Figure 4d, taken at the halfway point of the voltage shift. The shift clearly occurs at the point of grain coalescence with significant populations of both free-standing and coalesced grains in evidence. What is less evident is why the magnitude of ΔE_{cell} correlates closely to the cell performance. The superior performance of cells with lower ΔE_{cell} results from both a higher photocurrent and a better fill factor, as shown in Figure 4b. Both observations suggest that recombination in the diode is reduced at lower ΔE_{cell} while the open circuit voltage (V_{oc}) remains limited by factors independent of of charge recombination; most likely the Fermi level offsets intrinsic to the $Cu_2O|ZnO$ heterostructure.

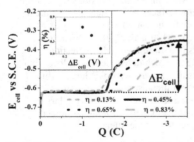

Figure 5a: Electrochemical cell potential (E_{cell}) measured on the Saturated Calomel Electrode (SCE) during Cu_2O deposition plotted as a function of the charge passed through the cell. Inset: resulting cell efficiency as a function of ΔE_{cell}.

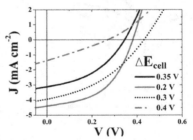

Figure 5b: Current-voltage curves for $Cu_2O|ZnO$ cells exhibiting different values for ΔE_{cell} during deposition. Cells are measured under illumination from AM 1.5 solar spectrum.

Two-step Cu_2O Deposition

Based on the work discussed above a method of electro-oxidation is introduced to combat series resistance experienced at lower pH. It is well known that the Cu_2O charge carrier density is governed by the actual ratio of Cu to O in the film; excess O, exhibited in the form of Cu vacancies, give rise to the p-type conduction commonly seen in Cu_2O[20–22]. A positive bias, typically resulting in CuO growth is applied to the as-grown Cu_2O film immersed in the growth bath with the aim of increasing the oxygen content within the film, resulting in a higher charge carrier density and lower series resistance.

Electro-oxidation was found to have unpredictable effects. Films for both diodes and solar cells were deposited under negative bias at 1 mA cm^{-2} for 50 minutes at a pH of 12.1 before being positively biased for 16 minutes at 3.5 mA cm^{-2}. Figure 6a shows solar cell performance efficiency marginally below that of the optimized one-step deposition. Diodes fabricated using the electro-oxidation step mostly failed to rectify with the exception of certain samples which exhibited notably improved built-in voltages, as shown in Figure 6b. Built-in voltages (V_{bi}) of approximately 0.6 V were determined by capacitance-voltage measurements (not shown) based on the Shockley ideal diode equation. To the best of the authors' knowledge, this is only the second instance of similarly high V_{bi}. Izaki *et al.* have achieved impressive cell efficiencies without deviating beyond the simple $Cu_2O|ZnO$ planar architecture with V_{bi} nearing 0.6 V[9]. They claimed a fine-tuned J_{dep} as being responsible for improved performance which has not been observed here. The inconsistency with this work would suggest that it is equally not explicable simply by the electro-oxidation process alone. The presence of unwanted phases leading to serendipitous rise in the V_{bi} can be precluded as an explanation as x-ray diffraction (not shown) shows only Cu_2O and ZnO phases. Likewise, microscope images show little deviation from as-deposited grains in Figure 4a. Apart from outright performance, the only sign of the electro-oxidation process affecting the as-deposited film is a distinct shift in color from deep red to a yellow brown. Work is ongoing to further quantify the effect on Cu_2O stoichiometry of the electro-oxidation process and it is the authors' opinion that any tentative explanation as to the origins of the high V_{bi} are held until such analysis can be completed.

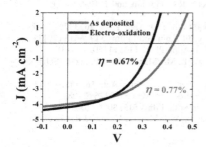
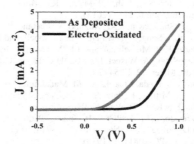

Figure 6a: Current-voltage curves for Cu2O|ZnO cells deposited galvanostatically at -1 mA cm-2 at pH of 12.05, with and without electro-oxidation at 3.5 mA cm-2.

Figure 6b: Dark Current-voltage curve for Cu2O|ZnO diodes with and without electro-oxidation.

CONCLUSIONS

The performance of $Cu_2O|ZnO$ solar cells fabricated by electrodeposition of Cu_2O is shown to be considerably more sensitive to the pH of the copper sulfate bath than has been previously reported. An optimal pH window of 12.0 – 12.1 is observed and is largely insensitive to the deposition current under galvanostatic deposition. A novel 2-step deposition is introduced in order to improve Cu_2O film quality through partial electro-oxidation of the as-deposited Cu_2O film. The electro-oxidation step is found to drastically, albeit inconsistently, improve the built-in voltage of the $Cu_2O|ZnO$ diode to 0.6 V. The challenge remains to fully explain the origins of this improvement and translate it into improved photovoltaic performance of the $Cu_2O|ZnO$ system.

ACKNOWLEDGMENTS

The research leading to these results has received funding from the European Community's 7th Framework Programme under grant agreement NMP3-LA-2010-246334. Financial support of the European Commission is therefore gratefully acknowledged.. SH would like to acknowledge the EPSRC University of Cambridge Doctoral Training Centre for Nanoscience and Girton College Cambridge for financial assistance.

REFERENCES

[1] Y. Zhou and J.A. Switzer, Scr. Mater. **38**, 1731 (2000).
[2] C. Wadia, a P. Alivisatos, and D.M. Kammen, Environ. Sci. Technol. **43**, 2072 (2009).
[3] A. Photovoltaics, S. Ru, A.Y. Anderson, H. Barad, B. Kupfer, Y. Bouhadana, and A. Zaban, (2012).
[4] K.P. Musselman, Nanostructured Solar Cells, University of Cambridge, 2010.
[5] B.K. Meyer, A. Polity, D. Reppin, M. Becker, P. Hering, P.J. Klar, T. Sander, C. Reindl, J. Benz, M. Eickhoff, C. Heiliger, M. Heinemann, C. Müller, and C. Ronning, Phys. Status Solidi **249**, 1487 (2012).
[6] B. Rai, Sol. Cells **25**, 265 (1988).
[7] J. Herion, E.A. Niekisch, and G. Scharl, Sol. Energy Mater. **4**, 101 (1980).
[8] Y.S. Lee, J. Heo, S.C. Siah, J.P. Mailoa, R.E. Brandt, S.B. Kim, R.G. Gordon, and T. Buonassisi, Energy Environ. Sci. **6**, 2112 (2013).
[9] M. Izaki, T. Shinagawa, K.-T. Mizuno, Y. Ida, and A. Tasaka, J. Phys. D. Appl. Phys. **40**, 3326 (2007).
[10] S. Jeong, S.H. Song, K. Nagaich, S. a. Campbell, and E.S. Aydil, Thin Solid Films **519**, 6613 (2011).
[11] A.T. Marin, K.P. Musselman, and J.L. MacManus-Driscoll, J. Appl. Phys. **113**, 144502 (2013).
[12] K.P. Musselman, A. Wisnet, D.C. Iza, H.C. Hesse, C. Scheu, J.L. MacManus-Driscoll, and L. Schmidt-Mende, Adv. Mater. **22**, E254 (2010).
[13] W. Septina, S. Ikeda, M.A. Khan, M. Matsumura, and L.M. Peter, Electrochim. Acta **56**, 4882 (2011).
[14] S. Jeong, A. Mittiga, E. Salza, A. Masci, and S. Passerini, Electrochim. Acta **53**, 2226 (2008).
[15] L.C. Wang, N.R. de Tacconi, K. Rajeshwar, and M. Tao, Thin Solid Films **515**, 3090 (2007).
[16] L.G. Gancedo, G.M. Ashley, X.B. Zhao, J. Pedrós, A. J. Flewitt, W.I. Milne, J.K. Luo, J.R. Lu, C.J.B. Ford, and D. Zhang, Int. J. Nanomanuf. **7**, 371 (2011).
[17] J. Stareck, USPO 2081121 (1935).
[18] M. Pourbaix, *Atlas of Electrochemical Equilibria*, 2nd Englis (Natl Assn of Corrosion, 1974).
[19] A. Osherov, C. Zhu, and M.J. Panzer, Chem. Mater. **25**, 692 (2013).
[20] D. Scanlon, B. Morgan, G. Watson, and A. Walsh, Phys. Rev. Lett. **103**, 1 (2009).
[21] H. Raebiger, S. Lany, and A. Zunger, Phys. Rev. B **76**, 1 (2007).
[22] G.K. Paul, Y. Nawa, H. Sato, T. Sakurai, and K. Akimoto, Appl. Phys. Lett. **88**, 141901 (2006).

Mater. Res. Soc. Symp. Proc. Vol. 1675 © 2014 Materials Research Society
DOI: 10.1557/opl.2014.848

Antibody immobilization for ZnO nanowire based biosensor application

Ankur Gupta, Monalisha Nayak, Deepak Singh, Shantanu Bhattacharya
Department of Mechanical Engineering,
Indian Institute of Technology Kanpur, U.P. 208016, India

ABSTRACT

Due to the high surface area and good bio-compatibility of nano structured ZnO, it finds good utility in biosensor applications. In this work we have fabricated highly dense ZnO nano bundles with the assistance of self assembled poly methylsilisesquoxane (PMSSQ) matrix which has been realized in a carpet like configuration with implanted ZnO nano-seeds. Such high aspect ratio structures (~50) with carpet like layout have been realized for the first time using solution chemistry. Nanoparticles of PMMSQ are mixed with a nano-assembler Poly-propylene glycol (PPG) and Zinc Oxide nanoseeds (5-15 nm). The PPG acts by assembling the PMSSQ nanoparticles and evaporates from this film thus creating the highly porous nano-assembly of PMMSQ nanoparticles with implanted Zinc Oxide seeds. Nano-wire bundles with a high overall surface roughness are grown over this template by a daylong incubation of an aqueous solution of hexamethylene tetra amine and Zinc nitrate. Characterization of the fabricated structures has been extensively performed using FESEM, EDAX, and XRD. We envision these films to have potential of highly dense immobilization platforms for antibodies in immunosensors. The principle advantage in our case is a high aspect ratio of the nano-bundles and a high level of roughness in overall surface topology of the carpet outgrowing the zinc-oxide nanowire bundles. Antibody immobilization has been performed by modifying the surface with protein-G followed by Goat anti salmonella antibody. Antibody activity has been characterized by using 3D profiler, Bio-Rad Protein assay and UV-Visible spectrophotometer.

KEYWORDS: Composition & Microstructure /Features /morphology, Performance /Functionality /sensor, Synthesis & Processing/Chemical Reaction/sol-gel.

INTRODUCTION

High surface area nanostructured film offers greater sites for biological entities loading capacity, which enhances the sensitivity of biosensor. In order to achieve this, highly oriented ZnO nanowires can be better tool for improving Biosensing performance. The most exigent step in the construction of a biosensor is the immobilization of biomolecules over the surface of the nanowires without loss of biological activity. Antibody immobilization over a substrate/film is very crucial for immuno-sensors which find enormous applications in various fields including clinical and pharmaceutical chemistry, immune-affinity purification of proteins, enzyme reactors as well as in environmental analysis [1-3]. Because of possessing inherently high surface area, good bio-compatibility and non-toxicity, nano structured ZnO film has great potential for biosensor applications. [4]. The diameter of nanostructured material is usually akin to the size of the biomolecule, which instinctively makes them outstanding transducers for producing a signal. Biosensors proffer some advantages over standard antibody detection methods (i.e., immunoprecipitation, enzyme-linked immunosorbent assays, magnetic immunoassay), such as, rapid detection, low production cost, portability, and ease of mass manufacturing [5]. There exist several immobilization strategies over solid substrate such as spontaneous and non-specific passive adsorption of a protein to synthetic surface (ELISA), covalent binding to the functional group of a surface, immunochemical immobilization, other techniques, i.e. non-adsorptive and non-covalent binding. The immobilization strategy may cause loss of protein activity, which may result into random orientations and structural deformations induced. In order to maintain the biological activity of the immobilized biological species, these should be attached to film without influencing the active sites. Out of various immobilization strategy explored, mostly are based on the physicochemical and chemical properties of the surface and bio molecule to be immobilized.

EXPERIMENTS

Polymers used for film fabrication are Propylene Glycol Methyl Ether Acetate (PGMEA) from sigma Aldrich) (CAS: 484431-4L), Poly propylene glycol, (PPG),(SARC CAQ 5322-69-4, molecular weight: 20,000 g/mol) and Polymethylsilisesquoxane (PMSSQ) (SGR650F, Techmiglon) are mixed and the zinc oxide (ZnO) particles using Zinc acetate dehydrate (Source: SDFCL, India) and isopropyl alcohol (Source: SDFCL) dispersed in the solution with assistance of ultra-sonication. Anti salmonella antibody was purchased from Abcam. Recombinant Protein G (lyophilized) was obtained from Sigma Aldrich, India. Bio-Rad protein assay dye reagent concentrate, IgG standard and BSA standards were purchased from Bio-Rad Laboratories (Rockville Center, NY).

Fabrication procedure for ZnO nano wire

Firstly, We prepare ZnO nano particles for dispersing it in the polymeric system made with PMSSQ and PPG. Zinc Acetate dihydrate (concentrations used 0.01M) in Isopropyl alcohol is well ultra-sonicated which converts the solution into lacteous colour. This solution is then heated at 200°C for a few seconds. At 200°C, the Zinc acetate crystals decompose to form ZnO nano particles.

Silicon (100) p-type substrates (Source: Logistic inc. NY) were appropriately cleaned with acetone, methanol, de-ionized (DI) water respectively. After proper cleaning, 40% Hydrofluoric acid (Source: SDFCL, India) mixed in a 1:5 ratio with DI water for passivating the silicon substrate resulting in a highly hydrophobic surface. Afterwards, the mixture of PMSSQ, PPG and ZnO nano particles in PGMEA is prepared and spun coated on the treated silicon substrates as thin films of thicknesses upto ~ 1 micron and is thereby heated to achieve a porous template holding the ZnO nano-seeds. The vertical structures are obtained on this template by putting the substrate upside down for 24 hours over a solution made up of Zinc Nitrate [$Zn(NO_3)_2.6H_2O$] (Source: Merck Specialities Pvt Ltd) and hexamethylenetetramine [$C_6H_6N_4$] (Source: Merck Specialites pvt ltd). The solution vapor transports the ZnO over the nano-seeds thus resulting in vertical ZnO nanostructures [6].The natural placement of the nano-seeds by the dispersion process is characterized by performing elemental mapping on the films using a field emission Scanning Electron Microscope (FESEM) (Zeiss Supra 40V,Germany). Further characterization of the nanostructures using X-ray diffraction(XRD) (Seift X Ray generator (ISO Debyeflex 2002,Germany)) with Cu-Kα source radiation (having wavelength 1.54 Å) for recording the vertical growth.

Immobilization procedure

Sodium acetate buffer was prepared by mixing 3.9 ml of 1M acetic acid properly with 6.10 ml of 1 M sodium acetate in 90 ml of DI water. We follow the procedure for immobilization as suggested by Babacan et al. (2000). Because of high affinity of protein G than protein A, we have used protein G in our case. Goat anti-salmonella is chosen for immobilization purpose, this recombinant protein G binds to Fc region of goat IgG. Firstly 10 mg/ml of protein G recombinant is prepared in phosphate buffer .Another Sodium acetate buffer is prepared as follows. 3.9 ml of 1ml acetic acid is mixed with 6.09 ml of 1M sodium acetate in 90 ml of DI water (pH of the solution is ~5). Stock solution of antibody is prepared by mixing 10mg/ml anti-salmonella in glycerol. Further working solution is made in PBS (Phosphate buffered saline). Final antibody solution is 10µl /ml of PBS. 10µl of the protein G solution is dispersed on the ZnO film. After dispersion, 10µl of 0.1 M sodium acetate buffer is poured over the film. 15µl of the antibody solution is applied over the protein treated surface.

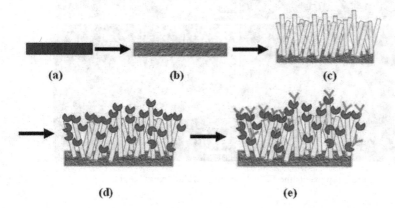

Figure 1 shows the schematic of the antibody immobilization over ZnO nano wires. (a) is Si substrate ,(b) is Si substrate with ZnO nano particles embedded polymeric film, (c) shows the ZnO nano wires grown over polymeric matrix, (d) Protein-G immobilized over the film and (e) shows the Antibody treatment over protein-G immobilized ZnO nano wire bundles

RESULTS AND DISCUSSION

The figure 2 represents the FESEM images of ZnO nano structures taken after a 24hour treatment in aqueous solution with precursor molar concentration ratio of 1:5. Figure 3 (a) is the XRD plot depicting the Plot clearly suggesting the wurtzite crystal lattice. Figure shows a dominance of (002) peak signifying high density and number of vertically oriented nanowires over the seeded substrate. Figure 3 (b) shows the presence of number of elements and their percentage amount in the film. Figure 4 (a) is the 3-D profiler image of the ZnO nano structures, figure 4 (b) is the image of protein-g immobilized film, figure (c) is the image of the anti-salmonella immobilized over the ZnO film. As can be clearly seen, roughness is decreased as the immobilization takes place which is due to accumulation of protein G and antibody solution in the vicinity of the film.

Figure 2 represents the FESEM images of ZnO nano bundles.

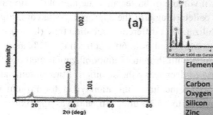

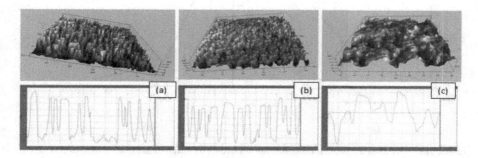

Element	Weight %	Weight % σ	Atomic %
Carbon	4.485	0.625	13.144
Oxygen	18.416	0.592	40.522
Silicon	6.735	0.348	8.441
Zinc	70.364	0.779	37.893

Figure 3: (a) XRD analysis of as deposited ZnO nanostructures over seeded Si-substrate, (b) shows the energy dispersive X ray (EDAX) analysis clarifying the presence of elements.

Figure 4 shows the 3D profiler results: (a) ZnO nano structures with extreme rough brush bed (b) protein-G based ZnO film (c) Antibody immobilized over ZnO film representing decreased roughness plot.

Calculations for Immobilization:

Percentage immobilization was calculated with following equation: $\frac{(P-Q)}{P} \times 100 = Protein\ immobilized$, Where P is the protein (μg) poured over the substrate and Q is the protein which could not bound to the substrate. Bio red Protein assay was used to determine the protein concentrations. The protein concentrations were determined by the micro Bio-Rad Protein Assay (with the help of Bio-Rad Protein Assay Manual). The protein concentrations in the antibody solution and Protein G solution were determined. The standard curves were obtained by preparing concentrations of standard IgG solutions for antibody binding and BSA standard for Protein G binding. Protein G is a directed immobilization method which is used here because of its natural affinity towards the Fc region present in the IgG molecules. This does not block the active sites of the antibodies for analyte binding [7,8]. The unbound protein was determined by rinsing the substrate with a known volume of buffer after the immobilization step and analyzing the protein content in the buffer collected against the appropriate standard. Fig. 5 shows the curve for percentage antibody immobilization at different incubation time (hours) with Protein G methods. Protein G–ZnO film binding was found from 60% at 0.5 h to ~77% after 3 h of incubation which was the highest yield obtained with Protein-G immobilization method. The decrease in immobilization was possibly because of decreasing the stability of the proteins after extended times at ambient temperature. At 27°C, the immobilization yield for Protein G–antibody binding was 52% after 0.5 h and decreased to ~47% after 1 h, and was approximately stable for another hour.

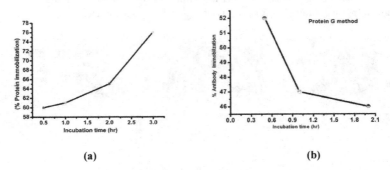

(a)   (b)

Figure 5(a) represents the protein immobilization yielding at different incubation time (0.5, 1, 2, 3 hrs.). Figure (b) represents the antibody immobilization with incubation time at temperature 27 °C.

CONCLUSION

ZnO nano bundles have much potential to be used for biosensing application. High aspect ZnO nano bundles have been grown successfully on the solid platform. Afterwards, Protein G method is adopted for antibody immobilization over the film. % Antibody immobilization achieved is ~52% at incubation time of 0.5 hour.

ACKNOWLEDGEMENT

Authors greatly acknowledge the Department of Science and Technology (DST), Government of India and Dean of Resource Alumni (DRPG), IIT Kanpur for providing financial support for attending the MRS conference. Authors also acknowledge Department of Biotechnology (DBT), India for providing funding to conduct this research. Authors would also like to thank Nano science centre, IIT Kanpur for providing all characterization facility for this work.

REFERENCES

1. Fransconi, M., Mazzei, F., Ferri, T., Anal. Bioanal. Chem. 2010, 398(4), 1545-1564.
2. Gooding, J.J., Hibbert, D.B.,Trends in Analytical Chem.,1999, 18(8), 525-533.
3. Wong, L.S., Khan, F., Mickfield, J., Chem. Rev.,2009, 109(9), 4025-4053.
4. Ramirez-Vick, J.E., J. Biosens. Bioelectron.,2012, 3(2), e109.
5. Strehlitz, B., Nikolaus, N., Stoltenburg, R., Sensors,2008, 8, 4296–4307.
6. Ankur Gupta, S. S. Pandey, Monalisha Nayak, Arnab Maity, Subhashish Basu Majumder, Shantanu Bhattacharya, "RSC adv.", 2014,4,7476-7482.
7. S. Babacan, P. Pivarnik, S. Letcher, A.G. Rand, Biosensors & Bioelectronics 15 (2000) 615–621
8. Deshpande, S.S., Antibodies: Biochemistry, structure, and function. In: Enzyme Immunoassays: From Concept to Product Development. Chapman and Hill, NY,1996, pp. 24–51.

Mater. Res. Soc. Symp. Proc. Vol. 1675 © 2014 Materials Research Society
DOI: 10.1557/opl.2014.836

Thermal stability of post-growth-annealed Ga-doped MgZnO films grown by the RF sputtering method

Kuang-Po Hsueh,[1] * Po-Wei Cheng,[1] Wen-Yen Lin,[1] Hsien-Chin Chiu,[2] Hsiang-Chun Wang,[2] Jinn-Kong Sheu,[3] and Yu-Hsiang Yeh[3]

[1] Department of Electronics Engineering, Vanung University, Chung-Li 32061, Taiwan
[2] Department of Electronics Engineering, Chang Gung University, Tao-Yuan 33302, Taiwan
[3] Institute of Electro-Optical Science and Engineering, National Cheng Kung University, Tainan 70101, Taiwan
* Electronic mail: kphsueh@mail.vnu.edu.tw

ABSTRACT

A radio-frequency magnetron sputtering technique and subsequent rapid thermal annealing (RTA) at 600, 700, 800, and 900 °C were implemented to grow high-quality Ga-doped $Mg_xZn_{1-x}O$ (GMZO) epi-layers. The GMZO films were deposited using a radio-frequency magnetron sputtering system and a 4 inch $ZnO/MgO/Ga_2O_3$ (75/20/5 wt %) target. The Hall results, X-ray diffraction (XRD), and transmittance were determined and are reported in this paper. The Hall results indicated that the increase in mobility was likely caused by the improved crystallization in the GMZO films after thermal annealing. The XRD results revealed that $Mg_xZn_{1-x}O$ (111) and MgO_2 (200) peaks were obtained in the GMZO films. The absorption edges of the as-grown and annealed GMZO films shifted toward the short wavelength of 373 nm at a transmittance of 90%. According to these results, GMZO films are feasible for forming transparent contact layers for near-ultraviolet light-emitting diodes.

INTRODUCTION

The $Mg_xZn_{1-x}O$ (MZO) alloy, has received increasing attention. Because of its highly tunable bandgap, from 3.37 eV of wurtzite (WZ) ZnO to 7.8 eV of rock-salt (RS) MgO, it is a promising candidate for deep ultraviolet (UV) optoelectronics devices and high-power electronics. [1], [2] Therefore, fabricating stable $Mg_xZn_{1-x}O$ films to widen the range of usable wavelengths and improve the efficiency of quantum confinement structures would be a major development in bandgap engineering.

In this study, the effects of thermal annealing on Ga-doped $Mg_xZn_{1-x}O$ (GMZO) films were investigated. GMZO films were deposited using radio-frequency (rf) magnetron sputtering followed by annealing at 600, 700, 800, and 900 °C in a nitrogen ambient for 60 s. Hall measurements were used to characterize the GMZO films. Additionally, the films characterization was evaluated using X-ray diffraction (XRD) with Cu-Kα 0.154 nm line as the radiation source (Bruker's XAS D5005). Transmission spectra were measured to investigate the optical properties of the GMZO films.

EXPERIMENT

The GMZO films used in this study were all grown at room temperature using the rf sputtering method on c-face sapphire substrates. The GMZO films were formed using a 4 inch ZnO/MgO/Ga$_2$O$_3$ (75/20/5 wt %) target. Before being loaded into a deposition chamber, the substrates were cleaned in an ultrasonic bath, using acetone, isopropyl alcohol, and deionized water for 5 minutes, respectively. The targets were pre-sputtered for 15 minutes before growth. Sputtering was carried out at an argon flow rate of 9 sccm, a pressure of 1.2×10^{-3} Torr, and a sputtering power of 150 W. The growth time was 90 minutes, and the chamber was cooled using a water-cooled chiller system during deposition. The thickness of the deposited GMZO film was 300 nm, as measured using a Veeco/Dektak 6M profiler. The GMZO samples were then annealed at 600, 700, 800, and 900 °C for 60 seconds each in a nitrogen ambient, in a rapid thermal annealing (RTA) system.

Table I. Summary of the Hall measurement on the Ga-doped Mg$_x$Zn$_{1-x}$O films annealing at different temperatures.

Temperature	Sheet Resistivity ohm/sq	Mobility cm^2/V-s	Concentration /cm^3
as-deposited	N.A.	N.A.	N.A.
600°C/N$_2$/60s (RTA600°C)	2.23×10^4	0.99	9.37×10^{18}
700°C/N$_2$/60s (RTA700°C)	3.82×10^4	0.34	1.58×10^{19}
800°C/N$_2$/60s (RTA800°C)	1.47×10^3	5.57	2.48×10^{19}
900°C/N$_2$/60s (RTA900°C)	1.55×10^3	11.18	1.20×10^{19}

DISCUSSION

Room temperature (RT) Hall measurements were determined to identify whether changes had occurred in the electrical properties of the GMZO films. Table I presents the dependence of the carrier concentration, electrical resistivity, and Hall mobility of the GMZO films on the sapphire substrate at room temperature. The as-deposited GMZO film did not show the Hall results because of its high resistivity. However, resistivity values of 2.23×10^4, 3.82×10^4, 1.47×10^3, and 1.55×10^3 ohm/square associated with electron concentrations of 9.37×10^{18}, 1.58×10^{19}, 2.48×10^{19}, and 1.20×10^{19}, cm^{-3} were obtained following annealing at 600, 700, 800 and 900 °C, respectively. The increase in mobility was likely caused by the improved crystallization in the GMZO films following thermal annealing.

Fig. 1 shows the XRD patterns of the n-type GMZO films before and after RTA. The pattern of the as-grown GMZO film exhibited a strong Mg$_x$Zn$_{1-x}$O (111) diffraction peak of 34.85 degrees coupled with MgO$_2$ (200) diffraction peaks of 37.65 degrees. The Mg$_x$Zn$_{1-x}$O (111) and MgO$_2$ (200) diffraction peaks of the annealed samples shifted to lower angles (larger lattice) at high annealing temperatures.

Fig. 2 presents the transmittance spectra of the GMZO films. The as-grown GMZO film in this study exhibited a high transparency with a transmittances over 95% in the visible region (360 - 700 nm) and a sharp absorption edge visible in the UV region (320 - 350 nm) because of the Mg content. The absorption edges of the as-grown GMZO films shifted toward a short wavelength of 320 nm at a transmittance of 80%. Furthermore, the transmittances of the GMZO films began decreasing at approximately 361 nm. The absorption edges of the annealed GMZO films changed to longer wavelengths at high annealing temperatures.

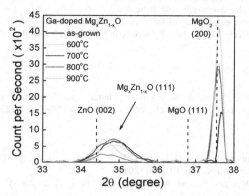

Fig. 1 XRD results of as-grown and annealed Ga-doped $Mg_xZn_{1-x}O$ films deposited on sapphire substrates.

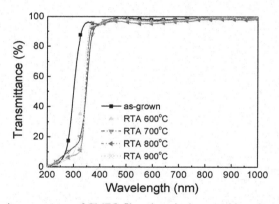

Fig. 2 The transmittance spectra of GMZO films deposited on sapphire substrates.

CONCLUSIONS

In this study, the thermal effects of GMZO films were investigated. GMZO films were deposited using an rf magnetron sputtering system and a $ZnO/MgO/Ga_2O_3$ (75/20/5 wt %) 4-inch target. The Hall results, XRD, and transparent performance were also measured. The XRD results indicated that the $Mg_xZn_{1-x}O$ (111) and MgO_2 (200) diffraction peaks of the annealed samples shifted to lower angles at high annealing temperatures. The as-grown GMZO film exhibited a high transparency with a transmittance over 95% in the visible region (360 - 700nm); its sharp absorption edge was visible in the UV region (323 - 350nm) because of the Mg content. The absorption edges of the as-grown and annealed GMZO films shifted toward the short wavelength of 373 nm at a transmittance of 90%. According to these findings, GMZO films are feasible for forming transparent contact layers for near-ultraviolet light-emitting diodes (LEDs).

ACKNOWLEDGMENTS

The authors would like to thank the National Science Council of the Republic of China, Taiwan, for financially supporting this research under Contract No. NSC 102-2221-E-238-007.

REFERENCES

1. M. D. Neumann, C. Cobet, N. Esser, B. Laumer, T. A. Wassner, M. Eickhoff, M. Feneberg, and R. Goldhahn, J. Appl. Phys. 110, 013520 (2011).
2. X. Chen, K. Ruan, G. Wu, and D. Bao, Appl. Phys. Lett. 93, 112112 (2008).

Mater. Res. Soc. Symp. Proc. Vol. 1675 © 2014 Materials Research Society
DOI: 10.1557/opl.2014.837

Emission Diversity of ZnO Nanocrystals with Different Growth Temperatures

E. Velázquez Lozada[1*], T. Torchynska[2], G. Camacho González[3]
[1]SEPI – ESIME – INSTITUTO POLITECNICO NACIONAL, México D. F. 07738, México.,
[2]ESFM – INSTITUTO POLITECNICO NACIONAL, México D. F. 07738, México.
[3]ESIME – INSTITUTO POLITECNICO NACIONAL, México D. F. 07738, México.

ABSTRACT

Scanning electronic microscopy (SEM), X ray diffraction (XRD) and photoluminescence (PL) have been applied to the study of structural and optical properties of ZnO nanocrystals prepared by the ultrasonic spray pyrolysis (USP) at different temperatures. The variation of temperatures and times at the growth of ZnO films permits modifying the ZnO phase from the amorphous to crystalline, to change the size of ZnO nanocrystals (NCs), as well as to vary their photoluminescence spectra. The study has revealed three types of PL bands in ZnO NCs: defect related emission, the near-band-edge (NBE) PL, related to the LO phonon replica of free exciton (FE) recombination, and its second-order diffraction peaks. The PL bands, related to the LO phonon replica of FE, and its second-order diffraction in the room temperature Pl spectrum testify on the high quality of ZnO films prepared by the USP technology.

INTRODUCTION

Nanocrystalline Zinc oxide (ZnO) with wide band gap energy nearly 3.37 eV, high exciton binding energy (60 meV at 300K) and easy way of nanostructure preparation has attracted great attention during the last two decades [1]. In addition to exceptional exciton properties, ZnO possesses a number of deep levels that emit in the whole visible range and, hence, can provide intrinsic "white" light emission. ZnO nanostructures are being investigated as promising candidates for different optoelectronic applications, such as: the non-linear optical devices [2], light-emitting devices [3-6], transparent electrodes for solar cells [7] and laser diodes [8], as well as for the excellent field emitters [9], electrochemical sensors and toxic gas sensors [10]. The control of the ZnO defect structure in these nanostructures is a necessary step in order to improve the device quality. Since the structural imperfection and defects generally deteriorate the exciton related recombination process, it is necessary to grow the high quality films for efficient light-emitting applications. The ultrasonic spray pyrolysis (USP) method offers many advantages such as easy compositional modifications; easy introducing the various functional groups, relatively low annealing temperatures and the possibility of coating deposition on a large area substrate. It will be interesting to study the optical emission of USP produced ZnO NCs doped by Ag in order to identify the best regimes for obtaining the bright emitting nanosystems.

EXPERIMENTAL DETAILS

ZnO:Ag thin solid films were prepared by the USP technique (Fig. 1) on the surface of soda-lime glass substrate for the two substrate temperatures (400 and 450 °C) and different deposition times (Table 1).

Table 1. Technological regimes and NC parameters from SEM images

Sample numbers	Growth temperature (°C)	Deposition time (min)	NR width (nm)
1	400	10	150-200
2	400	5	100-150
3	400	3	50-70
4	450	10	200-250
5	450	5	120-180
6	450	3	80-110

The deposition system presented in figure 1 includes a piezoelectric transducer operating at variable frequencies up to 1.2 MHz and the ultrasonic power of 120 W. ZnO:Ag thin solid films were deposited from a 0.4 M solution of zinc (II) acetate $[Zn(O_2CCH_3)_2]$ (Alfa), dissolved in a mix of deionized water, acetic acid $[CH_3CO_2H]$ (Baker), and methanol $[CH_3OH]$ (Baker) (100:100:800 volume proportion). Separately, a 0.2 M solution of silver nitrate $[Ag(NO_3)]$ (Baker) dissolved in a mix of deionized water and acetic acid $[CH_3CO_2H]$ (Baker) (1:1 volume proportion) was prepared, in order to be used as doping source. A constant [Ag]/[Zn] ratio of 2 at. % was applied at the ZnO Ag film preparation.

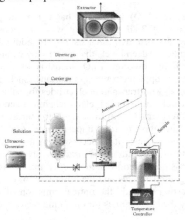

Figure 1. Schematic diagram of the experimental setup used for the deposition of the ZnO:Ag films by the ultrasonic spray pyrolysis method.

Concentration solution of Zinc solution precursor was 12.5 mM and Ag solution was 2.5 mM. The morphology of ZnO:Ag films has been studied by secondary electrons signals using the scanning electron microscopy (SEM) Dual Beam, FEI brand, model Quanta 3D FEG with field emission gun. PL spectra were at the excitation by a He-Cd laser with a wavelength of 325 nm and a beam power of 20 mW at 300K using a PL setup on a base of spectrometer SPEX500 described in [11, 12]. The crystal structure of ZnO:Ag films was investigated by the X-ray diffraction (XRD) using the diffractometer Modelo XPERT MRD with the Pixel detector, three axis goniometry and parallel collimator with the resolution of 0.0001 degree. XRD beam was from the Cu source, Kα1 line λ=1.5418 Å, 45 kV, 40 mA and the angles used were from 20° to 80° with a step size of 0.05° and step time of 100 s.

EXPERIMENTAL RESULTS AND DISCUSSION

SEM images of the typical ZnO:Ag NCs obtained at the deposition times of 3 and 10 min for two substrate temperatures are presented in figure 2. It is clear that the ZnO NCs have the hexagonal cross section and the road orientation along the c axis. The cross section of ZnO NCs increases with the temperature and durations of the UPS process (Table 1).

Figure 2. SEM images of the samples prepared at the substrate temperatures 400°C (a, b) and 450°C (c, d) and the durations of 10 (a, c) and 3 (b, d) min.

X-ray diffraction patterns of ZnO:Ag NCs obtained on the substrate with the temperature of 400°C and different deposition durations are shown in Fig. 3. At low deposition time (3min) the ZnO films showed an amorphous phase mainly with very small (100), (002) and (10 1) peaks. These peaks are the evidence of starting the conversion of an amorphous phase into a polycrystalline one. It was observed that with increasing the deposition duration from 3min to 10min, a set of new peaks appears, which correspond to the X- ray diffraction from the (100), (002), (101), (102), (110), (103) and (200) crystal planes (the angles 2θ equal to 31.770, 34.422, 36.253, 47.540, 56.604, 62.865 and 68.709 degrees), respectively, in the wurzite ZnO crystal structure with the hexagonal lattice parameters of a=3.2498A and c=5.2066A [13]. The (0 0 2) reflection peak is intense and sharper, as compared to the other peaks, indicating a preferential c-axis orientation of ZnO:Ag NCs. In X-ray diffraction patterns Ag dopant has not been identified because of USP technical is not appropriate to deposit Ag in ZnO films.

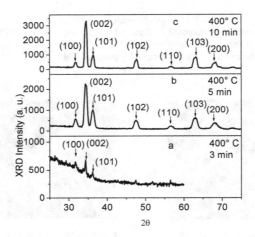

Figure 3. XRD diagrams of studied samples prepared on the substrate with T = 400° C at the deposition times of 3 (a), 5 (b) and 10 (c) min.

PL spectra of ZnO:Ag NCs are shown in figure 4. It is clear that the PL spectra are complex and can be represented as a superposition of elementary PL bands with the peaks in the spectral ranges: 3.14 eV (I), 2.00-2.70 eV (II) and 1.57 eV (III) (Fig.4). It is known that the UV-visible PL bands in ZnO owing to the near-band-edge (NBE) or exciton (I) and defect-related (II) recombination [14-17]. With increasing the substrate temperature the NBE related PL bands in the range I enlarged mainly, in comparison with the defect related (II) PL bands (Fig.4). At the same time the PL peak position of defect related PL bands shifts into the high energy (to 2.5 eV). With increasing the USP durations the intensity of defect related PL bands (II) raises mainly in comparison with those of NBE PL bands (Fig.4).

A great variety of luminescence bands in the UV and visible spectral ranges have been detected in the ZnO crystals [14]. The origin of these emissions has not been conclusively established. The NBE emission at 3.0-3.37 eV is attributed to the free (FE) or bound (BE) excitons, their LO phonon replicas, such as FE-1LO or FE-2LO, to optical transition between the free to bound states, such as the shallow donor and valence band, or to donor-acceptor pairs [14].

The blue PL band with the peak at 2.75-2.80 eV is attributed to Ag related defects [18], to Zn interstitials [19] or to donor-acceptor pairs including the shallow donor and oxygen vacancy [20]. The defect related green PL band in the spectral range 2.20-2.50 eV in ZnO is assigned ordinary to oxygen vacancies [19], Ag impurities [18] or surface defects [21]. The orange PL band with the peak at 2.02-2.10 eV was attributed earlier to oxygen interstitial atoms (2.02 eV) [22] or to the hydroxyl group (2.10eV) [23]. Taking into account that the PL intensity of 2.25 eV PL band increased with raising the USP duration the assumption that the corresponding defects are related to oxygen vacancies looks very reliable.

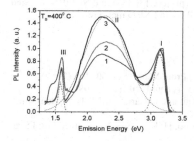 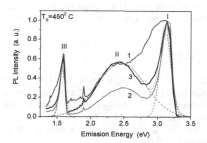

Figure 4a. PL spectra of samples prepared on the substrate with T = 400° C at the deposition times: 1-3 min, 2-5min, 3-10min.

Figure 4b. PL spectra of samples prepared on the substrate with T = 450° C at the deposition times: 1-3 min, 2-5min, 3-10min.

Studied ZnO films were doped by Ag and, therefore, have the acceptor type defects, Ag_{Zn}, which were formed when the Ag atoms substitute of Zn atoms in the ZnO crystal lattice. Emission of the acceptor type defects in ZnO has been studied intensively during the last two decades as well. Note, the broad PL bands with the peaks at 3.238 and 3.315 eV at 4.3K earlier have been observed in acceptor N doped MBE ZnO crystals [24]. These peaks were assigned to acceptor (N_o) bound exciton (3.315 eV), to donor-acceptor pair emission, involving N_o acceptor, or to the LO-phonon replica (3.238eV) of the donor BE line [24]. With temperature increasing the NBE intensity of acceptor bound exciton fallen down due to the dissociation of bound excitons and the FE band with it's LO replicas dominates in the PL spectrum. Due to the decrease in FE, the band of 2.20 2.50 eV (I I) decreases in PL intensity. Thus in our case the 3.14 eV PL band in the room temperature PL spectrum can be attributed to the LO phonon replica of FE emission. Note that the variation of PL intensity of the 1.57 eV PL band correlate with the intensity variation of the 3.14 eV PL band that permits to assign of the 1.57 eV PL band to the second-order diffraction peak of 3.14 eV PL band.

CONCLUSIONS

ZnO:Ag NCs with hexagonal structures have been successfully synthesized by the USP method. With increasing the substrate temperature at USP up to 450°C the PL intensity of NBE related emission bands has enlarged. The study has revealed three types of PL bands that in the room temperature PL spectrum related to the LO replicas of FE and its second-order diffraction peak, as well as the defect-related PL band, apparently, connected with oxygen vacancies. The PL band related to the LO phonon replicas of free exciton and its second-order diffraction in the PL spectra at room temperature testify on the high quality of the ZnO:Ag films prepared by USP.

ACKNOWLEDGMENTS

The authors would like to thank the CONACYT (project 130387) and SIP-IPN, Mexico, for the financial support, as well as the CNMN-IPN for SEM and XRD measurements.

REFERENCES

1. S.J. Pearton, D.P. Norton, K. Ip, Y.W. Heo, T. Steiner, Prog. Mater. Sci. **50**, 293 (2005).
2. M.H. Koch, P.Y. Timbrell, R.N. Lamb, Semicond. Sci. Technol. **10**, 1523 (1995).
3. K. Vanheusden, C.H. Seager, W.L. Wareen, D.R. Tallant, J. Caruso, M.J. Hampden-Smith, T.T. Kodas, J. Lumin.**75**, 11 (1997)
4. Z.K. Yang, P. Yu, G.L. Wong, M. Kawasaki, A. Ohtomo, H. Koinuma, Y. Segawa, Solid State Commun. **103**, 459 (1997).
5. N.H. Alvi, S.M. Usman Ali, S. Hussain, O. Nur and M. Willander Scripta Materialia **64**, 697 (2011).
6. M.H. Huang, S. Mao, H. Feick, Science **292,** 1897 (2001).
7] R. Scheer, T. Walter, H.W. Schock, M.L. Fearheiley, H.J. Lewerenz, Appl. Phys. Lett. **63,** 3294 (1993).
8. Y. Chen, D.M. Baghall, H. Koh, K. Park, K. Hiraga, Z. Zhu, T. Yao, J. Appl. Phys. **84,** 3912 (1998).
9. Y.B. Li, Y. Bando, D. Golberg, Appl. Phys. Lett. **84** , 3603 (2004).
10. J. Ding, T.J. McAvoy, R.E. Cavicchi, S. Semancik, Sens. Actuat. B **77** , 597 (2001).
11. M. Dybic, S. Ostapenko, T.V. Torchynska, E. Velazquez Lozada, Appl. Phys. Lett. **84** (25), 5165 (2004)
12] T. V. Torchynska, A.I. Diaz Cano, M. Dybic, S. Ostapenko, M. Mynbaeva, Physica B, Condensed Matter, **376-377,** 367 (2006)
13. PDF2 XRD database, card no. 36-1451.
14. A. B. Djuris, A.M.C. Ng, X.Y. Chen. Progress in Quantum Electronics 34. 191-259 (2010).
15. N.E. Korsunskaya, I.V. Markevich, T.V. Torchinskaya and M.K. Sheinkman, J. Phys. Chem. Solid. **43,** 475-479 (1982).
16. N.E. Korsunskaya, I.V. Markevich, T.V. Torchinskaya and M.K. Sheinkman, J. Phys. C. Solid St.Phys. **13**, 2975 -2982 (1980).
17. N.E. Korsunskaya, I.V. Markevich, T.V. Torchinskaya and M.K. Sheinkman, phys. stat. sol

(a), **60**, 565 -572 (1980).

18. M.A. Reshchikova, H. Morkoc, B. Nemeth, J. Nause, J. Xie, B. Hertog, A. Osinsky, Physica B, Condensed Matter, 401–402, 358–361 (2007).

19. M.K. Patra, K. Manzoor, M. Manoth, S.P. Vadera, N. Kumar, J. Lumin. 128 (2) 267–272 (2008).

20. D.H. Zhang, Z.Y. Xue, Q.P. Wang, J. Phys. D: Appl. Phys. 35 (21) 2837–2840 (2002).

21. A.B. Djurišic, W.C.H. Choy, V.A.L. Roy, Y.H. Leung, C.Y. Kwong. K.W. Cheah, T.K. Gundu Rao, W.K. Chan, H.F. Lui, C. Surya, Adv. Funct. Mater. 14 856-864 (2004).

22. X. Liu, X. Wu, H. Cao, R.P.H. Chang, J. Appl. Phys. 95 (6) 3141–3147 (2004).

23. J. Qiu, X. Li, W. He, S.-J. Park, H.-K. Kim, Y.-H. Hwang, J.-H. Lee, Y.-D. Kim, Nanotechnology 20 155603 (2009).

24. D.C. Look, D.C. Reynolds, C.W. Litton, R.L. Jones, D.B.Eason, G. Cantwell, Appl. Phys. Lett. **81**, 1830 (2002).

Mater. Res. Soc. Symp. Proc. Vol. 1675 © 2014 Materials Research Society
DOI: 10.1557/opl.2014.831

Synthesis and electrical characterization of ZnO and TiO2 thin films and their use as sensitive layer of pH field effect transistor sensors

Jessica C. Fernandes and Marcelo Mulato

Department of Physics, FFCLRP/University of São Paulo, Av. Bandeirantes, 3900 – Ribeirão Preto/São Paulo – Brazil

ABSTRACT

Oxide thin films of zinc and titanium materials were deposited by different deposition techniques, to be applied as sensitive layers of pH sensor – EGFET device. The deposition techniques tested were dip-coat, spin-coat, electrodeposition and spray-pyrolysis. The routine and the parameters of each technique were changed aiming optimized the procedures. The pHs buffer solutions tested ranged from 2 to 12. ZnO thin film shows sensitivity about 23 mV/pH, while TiO_2 thin films shows only 13.8 mV/pH. The final purpose of this study is to optimize the parameters for each deposition technique for both oxide materials.

INTRODUCTION

The interest to investigate metal oxide nanostructures thin films has been increasing recently, due to their many unique properties linked to the nanometer size of the particles. Besides, deposited oxides had a large and variety applications in sensors devices, fuel cells and microelectronic circuits fabrication [1,2]. The pH sensor can be applied in medical and industrial fields, so the development of alternative pH electrodes, as pH sensor based on field effect transistor [3], solid-state [4], and microelectrode among others, spurred considerable advantage [5]. Both zinc (ZnO) and titanium oxides (TiO_2) deposited as thin films has been largely applied as pH sensor due to their larger ranging in pH selectivity. The utilization of these materials as pH sensor has great interesting currently. The main purpose of the development of field effect transistor sensors and biosensors to be used in medicine field is the possibility of miniaturizing these devices.

Different deposition techniques can change the structural properties of thin film, as the grain sizes, the roughness, and the porosity. These characteristics can directly influences on the sensitivity of the film when it is applied as a sensing layer of an ionic sensor. The extended-gate field effect transistor (EGFET) device possibilities the electrical characterization of ionic sensors.

EXPERIMENTAL DETAILS

For ZnO thin films, four different deposition techniques (as dip-coat, spin-coat, spray-pyrolysis, and electrodeposition) were tested, while for TiO_2 thin films only dip and spin-coat deposition techniques were used. Both spray-pyrolysis and electrodeposition were not applied to TiO_2 thin films due to the viscosity of the precursor solution.

Precursor solution for both ZnO and TiO_2 were an optimization of different sol-gels routines found in the literature.

The final ZnO sol-gel routine used in all depositions techniques, except electrodeposition, was composed by 13.7 g of dihydrate zinc acetate diluted in 100 ml of ethanol, with four portion of 175 µL of lactic acid 85%, added in a range of 30 min each one. The solution was left in a reflux column for a period of 2 hours and 75° C. For electrodeposition the precursor solution was composed by 5mM of zinc chloride ($ZnCl_2$) and 0.1 M potassium chloride (KCl).

Dihydrate zinc acetate, $ZnCl_2$, KCl were obtained from Cinetica, while ethanol, and lactic acid 85% were obtained from FMaia and Synth, respectively.

The final TiO_2 sol-gel routine was a solution composed by 1.2 µL of titanium tetraisopropoxide (TTIP), 13,500 µL of isopropyl alcohol, 140 µL of Milli-Di water and 20 µL of hydrochloric acid (HCL). The solution was stirred at 50° C until the gel formation; approximately 15 min. Sol-gel was immediately used after preparation [6].

The both sol-gels were deposited in different substrates, as fluorine-doped tin oxide (FTO) for ZnO tests and indium-doped tin oxide (ITO) for TiO_2 tests.

ZnO deposition techniques

Different parameters were tested for all techniques, but the optimized parameters used were: for (a) dip-coat, the withdrawal velocity was 1 cm.min^{-1} and was deposited four layers. Before each layer a 200° C pre-heating during 10 m was made, and after all layers deposited a 1 hour annealing (540° C) was used; (b) spin-coat, the spin frequency was 840 rpm for a period of 20 s. Also, 4 layers were deposited, and the same pre-heating and annealing of dip-coat were made; (c) spray-pyrolysis, the substrate was kept with a constant temperature of about 360° C during the spray deposition. The distance between the spray and the substrate was 18 cm and a annealing was made about 1 h with a temperature of 540° C. An inert gas (nitrogen – N_2) was used during the deposition; and (d) electrodeposition, a silver reference electrode, a platinum counter-electrode, and the substrate were immersed on a zinc chloride and potassium chloride solution. Before the electrodeposition, the solution was bubbled with an inert gas (Nitrogen – N_2), and followed by the addition of 2.5 mM hydrogen peroxide (H_2O_2). Applied voltage range was -0.46/-0.37 V with a 0.01 V.s^{-1} scan rate. The cell was maintained with a constant temperature of 70° C during the whole deposition; 50 cycles were made [7].

TiO_2 deposition techniques

For TiO_2 thin films, the sol-gel was deposited by: (a) dip-coat, withdrawal velocity of 1 cm.min^{-1} and four layers were deposited. Before each layer a 100° C pre-heating during 10 m was made, and after all layers deposited a 1 hour annealing was made with a constant temperature of 400° C and; (b) spin-coat, the spin frequency was 2280 rpm for a period of 16 s. Also 4 layers were deposited, and the same pre-heating and annealing of dip-coat were used. These parameters can changes considerable the surface morphology of the samples, as can be found in literature [8]. So, these parameters have to be completely controlled.

EGFET electric characterization

The electric characterization was performed using an EGFET device. It consists of a silver reference electrode connected to the drain of a commercial MOSFET (CD4007, Texas Instruments), and the oxide thin films connected to the gate. The source terminal was grounded.

Both the silver reference electrode and the oxide thin film were dipped into different buffer solutions (PBS), with pHs in the range 2-12, purchased from Cinetica. The voltage between the drain and the source (V_{ds}) was kept constant at 5 V while the voltage between the gate and the source (V_{gs}) varies in a range 0-5 V. Each measurement gives a current value (I_{ds}) collected by an HP Data Acquisition 34970A system. The voltage was applied in the system using an Agilent E3646A Dual Output DC Power Supply. The Vgs is, in fact, the V_{ref} (voltage corresponded to the reference electrode) plus the voltage built up between the solution and the oxide thin film (dV).

DISCUSSION

Several parameters were tested, such as the numbers of layers, the pre-heating temperature and time, the annealing temperature, as well as the deposition techniques parameters. The presented results show the procedure that leads to better pH selectivity, i.e., the gate can sense each pH separately, and the comparison with the others deposition techniques tested.

ZnO thin films

Spray-pyrolysis deposition technique presented the most pH selectivity. The same numbers of layers and annealing temperature for the others techniques were maintained for comparison and the results are shown below. All films were deposited on FTO substrates.

Figure 1 presented the sensitivity in function of the pH for ZnO thin films deposited on FTO substrate using spray-pyrolysis. The sensitivity is determined using fixed I_{ds} values, and plotting the correspondent V_{gs} for each pH. The slope of the linear fit gives the sensitivity value.

The 23.2 mV/pH is a value below to the Nernstian sensitivity theoretical value for potentiometric sensors (59.2 mV/pH). Though the sensitivity value is below than the theoretical value, the films presented pH selectivity.

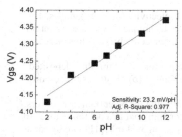

Figure 1. Gate-source voltage in function of pH with linear adjust for ZnO thin films deposited by spray-pyrolysis deposition technique on FTO substrate.

Figure 2 shows the sensitivity for thin films deposited by the other deposition techniques and for the FTO substrate as well. The other three depositions technique practically presented the same results of the FTO substrate. It indicates that the used parameters for the others techniques do not lead to a good adsorption of the precursor solution on the substrate, as presented for spray-pyrolysis. Besides, the selectivity only occurs for the alkaline pHs in a range 2-7 for electrodeposition and 2-6 for spin and dip-coats. For these ranges, the sensitivities had the respectively the values of 48.9, 39.5, and 40 mV/pH. The V_{gs} values of FTO were also plotted in function of each pH in order to comparison.

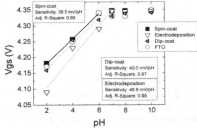

Figure 2. Gate-source voltage in function of pH with linear adjust for ZnO thin films deposited by: (a) spin-coat – solid square; (b) electrodeposition – up-side down triangles and; (c) dip-coat – left-side triangles; on FTO substrate - circles.

The deposition techniques changes considerable the final results, even trying to keep the same parameters as well as possible. Spray-pyrolysis shows a good adherence on substrate, due to the temperature used during the deposition, and presented good pH selectivity. For the others deposition techniques, the samples does not sense the difference between the alkaline pHs.

Changing the deposition technique the thickness, porosity and grain sizes also changes considerable, as already shown in literature [9,10]. To quantify these morphological features, further experiments will be performed, e.g., SEM and XRD.

TiO₂ thin films

Spin-coat deposition technique presented the most pH selectivity for TiO₂ samples. The same numbers of layers and annealing temperature for dip-coat technique was kept constant for comparison and the results are shown below, in figure 3 and 4, respectively. All films were deposited on ITO substrates.

Figure 3(a) presented the sensitivity in function of the pH for TiO₂ thin films deposited on ITO substrate using spin-coat deposition technique, while figure 3(b) illustrates the response for dip-coat technique.

Both values are also below to the Nernstian theoretical value for potentiometric sensors, but spin-coat presented good pH selectivity. The sensitivity for spin-coat in a pH range for 2-10 was about 13.8 mV/pH, while the sensitivity for dip-coat was about 39.4 mV/pH but only for a pH range 2-6. The film deposited by dip-coat does not show selectivity for alkaline pHs.

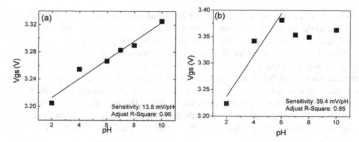

Figure 3. Gate-source voltage in function of pH for TiO$_2$ thin films deposited by: (a) spin-coat, with linear adjust for pH range 2-10 and; (b) dip-coat, with linear adjust for pH range 2-6.

The substrate of both films were ITO, wherein presented a sensitivity about 23 mV/pH but a selectivity for all pHs tested (figure 4). The dip-coat deposition leads to a lower selectivity in comparison with only the substrate, whilst spin-coat deposition leads to a lower sensitivity. Other parameters tested presented no pH selectivity even for FTO or glass substrates. New testes with different routine leads to films with sensitivities range between 23 and 55 mV/pH, with good adjust r-square, both in ITO and FTO substrates. The comparison between different deposition technique and the optimization of the parameters are currently been tested.

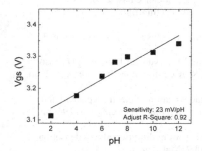

Figure 4. Gate-source voltage in function of pH for ITO substrate, with linear adjust for pH range 2-12.

Same as the ZnO thin films, the deposition techniques changes considerable the sensor response, due to the changes on sample morphology. Future tests will be made to quantify theses changes.

CONCLUSIONS

The deposition technique changes considerably the final structure characteristics of thin films. As a consequence, the sensitivity and the pH selectivity of films applied as the sensitive layer of a pH sensor-EGFET device are highly changed. For ZnO thin films, even trying to keep the same parameters for all techniques tested the final response changes significantly. Spray-pyrolysis presented a good adherence of sol-gel to the substrate. The others deposition technique does not presented any adherence. This characteristic is owed the temperature of the substrate during deposition. Other tests with the substrate temperature, not show in this paper, optimized the substrate temperature that shows the better adherence for each technique. TiO$_2$ thin films shows that the deposition technique besides changing the final response of the sensor also alters the sensitivity and the pH selectivity of the substrate. This study shows the importance of choice a deposition technique which best corresponds with the study purpose. A wrong choice can lead to a worst film with the same precursor solution and parameters that respond better for other deposition techniques.

ACKNOWLEDGMENTS

The work was supported by FAPESP, CAPES, and CNPq Brazilian agencies.

REFERENCES

[1] T. Larbi, B. Ouni, A. Boukachem, K. Boubaker, and M. Amlouk, "Electrical measurements of dielectric properties of molybdenum-doped zinc oxide thin films," *Mater. Sci. Semicond. Process.*, vol. 22, pp. 50–58, Jun. 2014.

[2] S. K. Mishra, S. Rani, and B. D. Gupta, "Surface plasmon resonance based fiber optic hydrogen sulphide gas sensor utilizing nickel oxide doped ITO thin film," *Sensors Actuators B Chem.*, vol. 195, pp. 215–222, May 2014.

[3] B. H. van der Schoot and P. Bergveld, "ISFET based enzyme sensors," *Biosensors*, vol. 3, no. 3, pp. 161–186, 1987.

[4] W. Olthuis, M. A. M. Robben, P. Bergveld, M. Bos, and W. E. van der Linden, "pH sensor properties of electrochemically grown iridium oxide," *Sensors Actuators B Chem.*, vol. 2, no. 4, pp. 247–256, Oct. 1990.

[5] T. Y. Kim and S. Yang, "Fabrication method and characterization of electrodeposited and heat-treated iridium oxide films for pH sensing," *Sensors Actuators B Chem.*, vol. 196, pp. 31–38, Jun. 2014.

[6] B. E. Yoldas, "Hydrolysis of titanium alkoxide and effects of hydrolytic polycondensation parameters," *J. Mater. Sci.*, vol. 21, no. 3, pp. 1087–1092, Mar. 1986.

[7] T. Pauporté and D. Lincot, "Hydrogen peroxide oxygen precursor for zinc oxide electrodeposition II—Mechanistic aspects," *J. Electroanal. Chem.*, vol. 517, no. 1–2, pp. 54–62, Dec. 2001.

[8] P-C. Yao, M-C-Lee, and J-L. Chiang. (2014). Annealing Effect of Sol-gel TiO2 Thin Film on pH-EGFET Sensor. Presented at Internacional Symposium on Computer, Consumer and Control.

[9] M-H. Habibi, and M. K. Sardashti. "Structure and Morphology of Nanostructured Zinc Oxide Thin Films Prepared by Dip- *vs.* Spin-Coating Methods" *J. Iran. Chem. Soc.*, vol. 5, no. 4, pp. 603-609, Dec. 2009.

[10] E. Nouri, et. al. "A comparative study of heat treatment temperature influence on the thickness of zirconia sol-gel thin films by three different techniques: SWE, SEM and AFM" *Surf. & Coat. Tech.*, vol. 206, pp. 3809-3815, 2012.

Mater. Res. Soc. Symp. Proc. Vol. 1675 © 2014 Materials Research Society
DOI: 10.1557/opl.2014.902

Comparative Study of the Antimicrobial Effect of Different Antibiotics Mixed with CuO, MgO and ZnO Nanoparticles in *Staphylococcus aureus*, *Pseudomonas aeruginosa* and *Escherichia coli* cultures

Raúl Alenó[1], Anthony López Collazo[1], Eulalia Medina [1], Lourdes Díaz Figueroa [1], José I. Ramírez[1] and Edmy J. Ferrer Torres[1]*.

[1]Department of Science and Technology, Inter American University of Puerto Rico, Ponce Campus, Mercedita, P.R. 00715. *Research Director, ejferrer@ponce.inter.edu

ABSTRACT

Due to the rapid advance of the emergence of resistant microorganisms to different antibiotics, there is a need to create new antimicrobial agents. It is possible that Nanotechnology has a great impact in this area since the nanoparticles can improve the antimicrobial effect of the antibiotics. In this study we used three different metal oxides nanoparticles, the MgO, ZnO and CuO. These nanoparticles were selected because their interactions leading to cell death and their optical properties. The aim of this study is to develop new methods that are more effective against resistance bacteria, developing antibacterial agents using different nanoparticles against *Escherichia coli* (ATCC 10536), *Pseudomonas aeruginosa* (ATCC 10145), and *Staphylococcus aureus* (ATCC BAA-1026). This study was conducted to evaluate the antibacterial effects of a combination of nanoparticles together with different concentrations of three antibiotics, Gentamicin, Cephalexin and Co-Trimoxazole. The results showed that some nanoparticles are effective to inhibit growth in these microorganisms by increasing the effectiveness of the antibiotic. Therefore, the present study indicates that the combination of the nanoparticles with antibiotics may be applicable as a new antimicrobial agent.

INTRODUCTION

The use of nanotechnology for the treatment of bacterial infections is an innovative strategy to enhance the effect of antibiotics. Recovery from bacterial infections is challenged due to the indiscriminate use and abuse of antibiotics [1]. Therefore, nanoparticles can be used to develop new treatments against resistant bacterial infections increasing the potential of antibiotics. This study presents antimicrobial activity of enhanced antibiotics mixed with CuO, MgO and ZnO nanoparticles. There is evidence indicating the antimicrobial effectiveness of ZnO, CuO and MgO [2,3,4]. Our goal is to develop more effective treatments against resistant bacterial infections combing antibiotics and nanoparticles. The antibiotics chosen to perform this study were Gentamicin, Cephalexin, and Co-Trimoxazole. Nanoparticles of MgO, CuO and ZnO were mixed with antibiotics, at different proportions and concentrations to determine their effectiveness.

EXPERIMENT

Nanoparticles were synthesized through the modification of reported methods in the literature [5,6]. CuO nanoparticles were synthesized dissolving 0.2497 g of copper (II) sulfate pentahydrate salt (CuSO$_4$.5H$_2$O) 0.01 M in 20 ml of PEG 0.02 M. The amount of ascorbic acid 0.02 M, added by titration, was 18ml. 500 μl of NaOH 0.1 M were added to the mixture followed by 20 μl of NaBH$_4$ 0.1 M. For the synthesis of MgO nanoparticles, 100 ml of magnesium sulfate solution (0.14M) was used with PVP as stabilizer and sonicated for 30 min. Fifty ml of NaOH solution (0.2M) was slowly added by titration. After NaOH was added, the mixture was sonicated for another 30 min. The precipitated was filtered and further on dehydrated at 200 °C for 2 h. At the dehydration phase, nanostructure magnesium oxide was formed. The final product was obtained in powder and colloid form. Nanoparticles of ZnO were synthesized with 6.25 ml zinc sulfate solution (0.10 M) and PVP as stabilizer. They were sonicated for 30 min. After the sonication process 12.5 ml of TMAH solution (0.28M) was slowly added by titration and the solution was sonicated for another 30 min. The filtered precipitate was dehydrated at 200 °C for 2 h. At the dehydration phase, nanostructure zinc oxide was formed. The final product was obtained in powder and solution form. CuO, MgO, and ZnO nanoparticles were characterized by Dynamic Light Scattering.

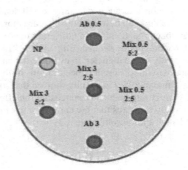

Figure 1. Method: Arrangement of the distribution of disks placed on the MHA medium after immersion in the solutions. Ab is for antibiotics, and NP is for nanoparticles. Mixes of proportions of ab:np are indicated.

Gentamicin, cephalexin and co-trimoxazole solutions were prepared at 0.5 and 3 mg/ml. Three bacteria were chosen to perform the study, *Staphylococcus aureus*, *Escherichia coli* and

*Pseudomonas aeruginosa.*The method used for the study was a modification of the Disk Diffusion Method of Kirby-Bauer. Antibiotics and nanoparticles were impregnated onto paper disks, and placed on a seeded plate of Mueller-Hinton Agar as shown in figure 1. Tryptic Soy Broth(TSB) and Mueller-Hinton Agar(MHA) media were prepared to perform the study. The incubation of *S. aureus, E. coli,* and *P. aeruginosa* was at 37 Celsius degree. The bacterial cultures were standardized at 0.5 Mc Farland Standard (1.5 x 108 CFU/ml). MHA plates were streaked with *S. aureus, E. coli,* and *P. aeruginosa.* Solutions for the disks were prepared as follows; nanoparticle solution, antibiotic solution (0.5, and 3mg/ml), nanoparticle/antibiotic mix solution 5:2 (125:50 µl) proportion, and nanoparticle/antibiotic mix solution 2:5 proportion. All the plates, not inverted, were incubated for 18 hours at 37 Celsius degrees. After the incubation period, the inhibition zones were measured in millimeters.

DISCUSSION

Table 1. Effect of mixtures of different proportions of antibiotics and nanoparticles during bacterial growth. .

NP	Bacteria	Antibiotic Inhibition Zones(mm) [Antibiotic] in mg/ml	Mixed Proportion Inhibition Zones Antibiotic:NP 5:2 in mm	Mixed Proportion Inhibition Zones Antibiotic:NP 2:5 in mm	Antibiotic Inhibition Zones(mm) [Antibiotic] in mg/ml	Mixed Proportion Inhibition Zones Antibiotic: NP 5:2 in mm	Mixed Proportion Inhibition Zones Antibiotic: NP 2:5 in mm	Antibiotic Inhibition Zones(mm) [Antibiotic] in mg/ml	Mixed Proportion Inhibition Zones Antibiotic: NP 5:2 in mm	Mixed Proportion Inhibition Zones Antibiotic: NP 2:5 in mm
CuO	S.aureus	GM [3] 12	13	11	CE [3] 10	9	0	CO [3] 30	30	26
		GM [0.5] 0	0	0	CE [0.5] 0	0	0	CO [0.5] 23	22	21
CuO	E.coli	GM [3] 32	34	31	CE [3] 0	0	0	CO [3] 0	0	0
		GM [0.5] 27	27	25	CE [0.5] 0	0	0	CO [0.5] 0	0	0
CuO	P.aeruginosa	GM 3 32	32	27	CE [3] 0	0	0	CO [3] 0	0	0
		GM [0.5] 22	22	17	CE [0.5] 0	0	0	CO [0.5] 0	0	0
MgO	E.coli	GM [3] 29	30	29	CE [3] 0	0	0	CO [3] 0	0	0
		GM [0.5] 25	23	22	CE [0.5] 0	0	0	CO [0.5] 0	0	0
MgO	P.aeruginosa	GM [3] 29	30	27	CE [3] 0	0	0	CO [3] 0	0	0
		GM [0.5] 22	21	15	CE [0.5] 0	0	0	CO [0.5] 0	0	0
MgO	S.aureus	GM [3] 29	29	27	CE [3] 0	0	0	CO [3] 31	31	27
		GM [0.5] 25	25	21	CE [0.5] 0	0	0	CO [0.5] 24	23	19
ZnO	S.aureus	GM [3] 13	11	0	CE [3] 9	0	0	CO [3] 25	22	12
		GM [0.5] 0	0	0	CE [0.5] 0	0	0	CO [0.5] 0	0	0
ZnO	E.coli	GM [3] 29	29	27	CE [3] 0	0	0	CO [3] 0	0	0
		GM [0.5] 24	23	20	CE [0.5] 0	0	0	CO [0.5] 0	0	0
ZnO	P.aeruginosa	GM [3] 33	31	27	CE [3] 0	0	0	CO [3] 0	0	0
		GM [0.5] 24	22	16	CE [0.5] 0	0	0	CO [0.5] 0	0	0

Inhibition zones were measured in millimeters as reported in table 1. Mixed nanoparticles of CuO and MgO with Gentamicin 3 mg/ml in a proportion of 5:2 inhibited more bacterial cells than the antibiotic by itself (figure 2). However, CuO and MgO nanoparticles decreased the antimicrobial activity of Cephalexin, and Co-trimoxazole when mixed . None of the antibiotics in a concentration of 0.5 mg/ml were enhanced by any of the nanoparticles.

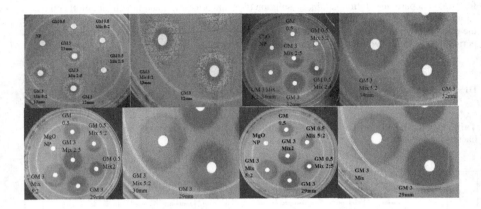

Figure 2. Inhibition zones of Gentamicin and nanoparticles in a seeded MHA medium. Top left, *S. aureus* with Gentamicin and CuO NP. Top right, *E. coli* with GM and CuO NP. Down left, *E. coli* with GM and MgO NP. Down right, *P. aeruginosa* with GM and MgO NP.

Our ZnO NP has a mean of 68 nm and did not showed a direct or indirect effect with nanoparticles against the three bacteria studied, but it's been reported that ZnO NP of diameters between 12 and 45 nm have an antimicrobial effect against bacteria [7]. CuO NP diameter mean was about 14 to 16 nm and MgO nanoparticles from 22 to 24 nm as shown in figure 3.

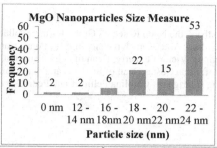

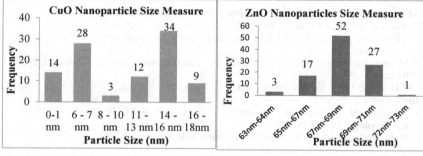

Figure 3. Dynamic Light Scattering spectre of nanoparticles. Top , Dynamic Light Scattering spectre of MgO NP. Down left, Dynamic Light Scattering spectre of CuO NP. Down right, Dynamic Light Scattering spectre of ZnO NP.

CONCLUSION

Copper Oxide (CuO) Nanoparticles increased the antimicrobial activity of Gentamicin at a concentration of 3mg/ml against *Staphylococcus aureus* and *Escherichia coli*. Magnesium Oxide(MgO) nanoparticles increased the antimicrobial activity of Gentamicin 3mg/ml against *Escherichia coli* and *Pseudomonas aeruginosa*. Mixtures proportions of 5:2 (antibiotic: nanoparticles) increased the antimicrobial activity of Gentamicin 3mg/ml against all three bacteria. Same proportions of 5:2 have better inhibition zones (inhibited more bacteria) compared to the Gentamicin 3mg/ml solution. Cephalexin, and Co-trimoxazole mixtures with CuO, ZnO and MgO nanoparticles decreased the antimicrobial activity of the antibiotics. ZnO nanoparticles was not effective enhancing the antimicrobial activity of antibiotics, and was not effective in the inhibition of bacteria.

ACKNOWLEDGMENTS

The authors would like to thank the Nano Research Group for their collaboration. Our research was supported by the Inter American University of P.R., Ponce Campus, Dr. Vilma E. Colon, Chancellor of our campus; Victor Feliberty, Dean of Administration, Dr. Jacqueline Alvarez, Dean of Studies, Prof. Edda Costas Dean of Students and our laboratory partners Marangelie Feliciano, Keishla Rodriguez , Carolina Pellicier and Pedro Rivera Pomales.

REFERENCES

1. P.T.A. Usha, S. Jose, and A.R. Nisha, *Veterinary World* 3, 138 (2010).
2. M. Sundrarajan, J.Suresh, and R.R. Gandhi, *Dig. J. Nanomater. Bios.* 7, 983 (2012).
3. J. P. Ruparelia, A.K. Chattergee, S.P. Duttagupta, and S. Mukherji, *Acta. Biomater.* 4, 707   (2008).
4. H. Meruvu, M. Vangalapati, S.C. Chippada, and S.R. Bammidi, *Rasayan J. Chem.* 4, 217 (2011).
5. T.M.D. Dang, T.T.T. Le, E.F. Blanc, and M.C. Dang, *Adv. Nat. Sci. Nanosci. Nanotechnol.* 2, 6 (2011).
6. S.D. Meenakshi, M. Rajarajan, S. Rajendran, Z.R. Kennedy, and G. Brindha, Elixir Nanotechnology. 50, 10618 (2012).
7. N. Padmavathy, R. Vijayaraghavan, Sci.Technol. Adv. Mater. 9, 7(2008).

Mater. Res. Soc. Symp. Proc. Vol. 1675 © 2014 Materials Research Society
DOI: 10.1557/opl.2014.852

Effect of Dopant Oxidation State and Annealing Atmosphere on the Functional Properties of Zinc Oxide-Based Nanocrystalline Powder and Thin Films

Miguel A. Santiago Rivera[1], Gina M. Montes Albino[2], and Oscar Perales Pérez[3]
[1] Dept. of Physics, University of Puerto Rico, Mayagüez PR 00681-9000, U.S.A.
[2] Mechanical Engineering Department, University of Puerto Rico, Mayagüez PR 00681-9000, U.S.A.
[3] Engineering Science and Materials Department, University of Puerto Rico, Mayagüez PR 00681-9000, U.S.A.

ABSTRACT

The effective incorporation of dopant species into ZnO host structure should induce changes in its physical and chemical properties enabling the establishment of novel multi-functional properties. Doping with transition metal ions and the subsequent exchange interaction between available spins of the magnetic species are expected to induce a ferromagnetic behavior. This ferromagnetic functionality will enable the application of this material in data storage and spintronics-based devices. The present research addresses the study of the effect of the oxidation state of Fe species and the influence of the annealing atmosphere on the structural and functional properties of nanocrystalline ZnO-based powders.

INTRODUCTION

Among various semiconductor oxides that are expected to become suitable platforms for multifunctional applications, zinc oxide (ZnO) is one of the most attractive alternatives. ZnO is a wide-bandgap semiconductor of the II-VI semiconductor group and exhibits several favorable properties, including good transparency, high electron mobility, wide band-gap, and strong room-temperature luminescence. These properties makes ZnO an excellent candidate for emerging applications such as transparent electrodes in liquid crystal displays, energy-saving or heat-protecting windows, thin-film transistors for electronics, and light-emitting diodes (LED's)[1,2]. The effective incorporation of dopant species into ZnO host structure should induce changes in its physical and chemical properties enabling the establishment of novel multi-functional properties. In the case of doping with transition metal ions, such as Fe, Mn and V ions, the subsequent exchange interaction from magnetic spins should induce a ferromagnetic behavior in the so-called ZnO-based diluted magnetic semiconductor. This ferromagnetic functionality will enable the application of this material in data storage and spintronics-based devices. On the other hand, among the acceptor impurities that affect ZnO, nitrogen (N) has been considered to be the most suitable p-type dopant due to atomic size and electronic structure considerations. The energy of the valence 2p states and the electronegativity of nitrogen are also the closest to those of the oxygen atom, particularly when compared with other column-V dopants. Several groups have reported the incorporation of N in ZnO, and many have claimed that N substitutes oxygen (O) species into the ZnO structure[1].

EXPERIMENTAL

Materials

Pure and Fe-doped ZnO powders and films were synthesized via a sol-gel technique. Zinc Acetate Dihydrate (Zn(CH$_3$COO)•2H$_2$O, 98%-101%), Iron (II) Chloride Tetrahydrate

(FeCl$_2$•4H$_2$O, 98%), and Iron (III) Nitrate Nonahydrate, Fe(NO$_3$)$_3$•9H$_2$O, 98%-101%, were used without further purification. Ethanol, (alcohol reagent, anhydrous denatured, 94-98%), and MEA (monoethanolamine) were used as solvent and viscosity-controlling additive, respectively. Quartz substrates (Quartz Plus Inc.) are ultrasonically cleaned in acetone, and wash out with ethanol and de-ionized water prior to spin- coating.

Synthesis of pure and Fe-doped nanocrystalline powders

A modified sol-gel approach was followed to synthesize well-crystallized ZnO-based nanocrystalline powders. Pure and Fe doped ZnO powders samples were synthesized by dissolving suitable amounts of Zn and dopant salts in ethanol and subsequent heating at 150°C for twelve hours to assure the complete removal of the solvent. The Fe concentration ranged between 1 and 10 at %. The obtained solid precursor thermally treated in air for 1 hour at 500°C to produce the desired oxide phase.

Materials characterization

The structure of produced powders was determined by means of X-ray diffraction (XRD) using a SIEMENS D-5000 unit with Cu-Kα radiation. Average crystallite sizes were estimated from the main diffraction peaks according to Scherrer's equation. The morphology and size were examined using a Transmission Electron Microscope (TEM) JEM-ARM200cF. Optical properties were measured using a UV-Vis Beckman Coulter DU 800 spectrophotometer and a Shimadzu RF-5301PC Spectro-fluorometer, respectively. Room-temperature M-H hysteresis loops of Fe- doped ZnO powders were recorded in a Lake Shore 7410 Vibrating Sample Magnetometer (VSM).

RESULTS AND DISCUSSION

X-ray diffraction analyses

Figure 1 and 2 shows the X-Ray diffraction patterns corresponding to Fe^{2+}- doped ZnO annealed in air and nitrogen at 500 °C, respectively. The diffraction peaks corresponding to the crystallographic planes in hexagonal wurtzite became evident. The same behavior was observed for those ZnO samples doped with Fe^{3+} (Figures 3 and 4). The absence of other isolated phases may suggest the possible incorporation of Fe ions within the ZnO host. The observed drop in the average crystallite size from 30 nm to 26 nm when annealed in air at 500°C can be attributed to the modification of the nucleation rate of the crystals due to the incorporation of the dopant species into the ZnO host structure.

Although not shown here, a slight but noticeable shift was observed for the diffraction peaks of the samples produced at different concentration of the dopant species. This shift could be related to a distortion in the ZnO unit cell due to the actual substitution of the iron ions in the zinc sites.

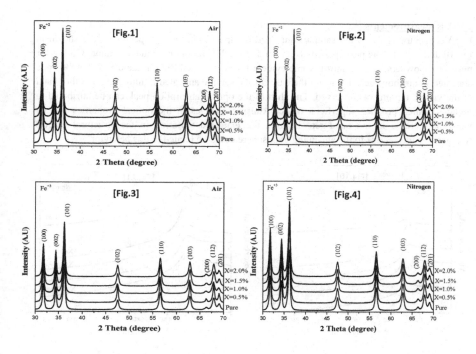

Fig.1-4 XRD patterns of ZnO samples doped with Fe(II) species in air and nitrogen (1 and 2) and Fe(III) in the same gas environments (3 and 4).

Transmission Electron Microscopy Analyses [TEM]

TEM images corresponding to Fe^{+2}(2 at.%)-ZnO are shown in figures 5 to 9. The nanometric nature of the particles was evidenced. The nanoparticles exhibited an approximate size of about 20 nm.

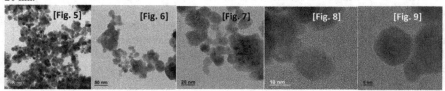

Fig.5-9 TEM images corresponding to Fe^{+2}(2 at.%)-ZnO nanoparticles

UV-Vis measurements

UV-Vis spectra of Fe^{+3}-ZnO samples annealed in air and nitrogen, respectively, are shown in Fig.10 and Fig.11. The principal emission band in the UV region is centered around 373 nm for pure ZnO powders. A blue shift became evident for Fe^{+3}-doped ZnO in the 0.5- 2.0 at % range annealed in air; however, it was not the case for the sample annealed in nitrogen. No other band in the visible region was observed. The shift of the optical absorption peak can be attributed to dopants, which produce superficial defects causing changes in the optical properties and an apparent increase in the band-gap. In turn, the expected variation in the structural defects (oxygen vacancies) due to the annealing atmosphere could be related to this behavior.

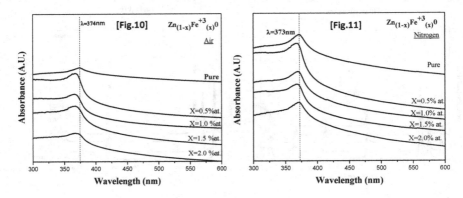

Fig.10-11. UV-Vis spectra of Fe^{+3}-doped ZnO annealed for one hour at 500°C in air and nitrogen, respectively.

PL Measurements

The photoluminescence of Fe^{+3}-doped ZnO nanoparticles annealed for one hour at 500°C in air and nitrogen were also measured. The corresponding spectra are shown in figures 12 and 13. The excitation wavelength was set at 345 nm. The main emission band was centered on 392 nm and is attributed to the radiative annihilation of excitons with a very short life-time. A quenching-by-concentration effect was observed in all the doped samples. This quenching can be attributed to the formation of trapping states, which in turn is an evidence of the incorporation of the dopant species in the oxide lattice[1,5,7]. It also can be attributed to a decrease in the average crystallite size corroborated by XRD.

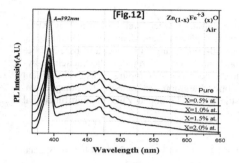 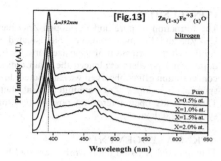

Fig.12-13. Room-temperature PL spectra of pure and Fe(III)-doped ZnO powders at different dopant concentrations and annealed at 500°C in air (12) or nitrogen (13).

Room-temperature M–H measurements

As figure 14 evidences a weak but noticeable ferromagnetism was observed in the Fe^{3+} (2 at.%)-doped ZnO sample. The coercivity value was 138 Oe and the maximum magnetization was 0.137emu/g. Observed ferromagnetism could be attributed to the contribution of magnetic moments of the dopant species actually incorporated in the host ZnO lattice.

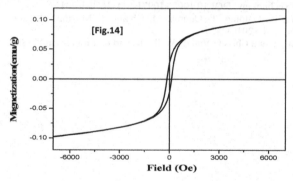

Fig.14 Room-temperature M-H loop of Fe^{3+}- doped ZnO powders (2 at.% Fe) annealed in air for one hour at 500°C.

CONCLUSIONS

The formation of pure and Fe- doped ZnO nanocrystalline powders were confirmed by XRD, UV-Vis and PL measurements. No isolated impurity phases were detected. TEM analyses confirmed the formation of nanosize particles in the 10-20 nm range. PL spectra of bare and Fe-doped ZnO powders evidenced the main emission band in the UV region. The quenching-by-concentration effect observed was attributed due to the formation of trapping states by the dopant species. Fe^{3+}-doped ZnO exhibited weak but noticeable room-temperature ferromagnetism. These results open interesting possibilities for the use of Fe-doped ZnO in multifunctional systems and devices.

ACKNOWLEDGEMENTS

This material is based upon work supported by the DOE Grant No. FG02-08ER46526. Special thanks to the Earth-X-ray Laboratory at the Department of Geology, UPRM.

REFERENCES

1. Boris Straumal, Svetlana G Protasova; Beilstein; Journal of Nanotechnology, **4,** (2013): 361
2. M Aslam and D Bahadur; Journal of Nanotechnology **23**, (2012): 115601
3. Dhiman Pooja; Research Journal of Recent Sciences, **1**(8), (2012): 48
4. G. MurtazaRai, M.A. Iqbal, Y.B. Xu and I.G.Will, Z.C. Huang; Journal of Magnetism and Magnetic Materials, (2011): 323
5. Anna Baranowska-Korczyc and Anna Reszka, Kamil Sobczak; Journal of Sol-Gel Science and Technology, DOI 10.1007/s10971-011-2650-1(2011)
6. A. Tsukazaki, A. Ohtomo, T. Onuma, M. Ohtani, T. Makino, M. Sumiya, K. Ohtani Nature Materials **4**, (2010), 42
7. Anderson Janotti and Chris G Van de Walle; Report on Progress in Physics, (2009):72

Mater. Res. Soc. Symp. Proc. Vol. 1675 © 2014 Materials Research Society
DOI: 10.1557/opl.2014.853

The Effect of oxygen defects on Activity of Au/ZnO Catalyst in Low Temperature Oxidation of Benzyl Alcohol

R. Shidpour[1,4],*, M. Vossoughi [1,2], and A.R. Simchi [1,3]

[1]Institute for Nanoscience & Nanotechnology, Sharif University of Technology, Tehran, Iran
[2]Chemical & Petroleum Engineering Department, Sharif University of Technology, Tehran, Iran
[3]Material science and engineering Department, Sharif University of Technology, Tehran, Iran
[4]Department of Chemistry, University of California, Riverside, USA

Abstract:

Gold nanoparticles supported on ZnO nanostructures were prepared through a simple chemical-thermal method and characterized by SEM, TEM, XRD and photo luminescence (PL) spectroscopy. Effect of annealing temperature on catalytic activity of these Au/ZnO nanocatalysts were investigated by aerobic oxidation of benzyl alcohol. The results indicated that the catalyst with ZnO nanowire support annealed at 300 °C exhibited more activity than Au/ZnO catalyst supported on ZnO nanoparticles annealed at 600 °C. The Au/ZnO-nanowire achieved to increase the benzaldehyde selectivity and yield to 93.7 % and 85.6 %, respectively, at 60 °C whereas in Au/ZnO-nanoparticle the benzaldehyde selectivity and yield to 85.1 % and 69.9 %, respectively at 80 °C. The XRD and PL spectroscopy revealed that the supports have interstitial zinc (Zn_i), oxygen vacancy (Vo^{-2}) defects definitely but there is no evidence for interstitial oxygen (O_i) and zinc vacancy (V_{Zn}) defects and single ionized charged oxygen vacancy (Vo^-).

1. Introduction

The oxidation of alcohols into their corresponding aldehydes and ketones is one of the most important reactions in both laboratory and industrial synthetic chemistry [1]. In recent years, the high catalytic activity of Au as a low temperature CO oxidation catalyst [2] has started intensive research in the use of Au nanoparticles for the liquid phase oxidation of alcohols [3-7]. In comparative studies of aerobic catalytic oxidation of alcohol using supported Pt, Pd, or Au as catalyst, Au has shown to have the highest selectivity and to be less prone to metal leaching, over-oxidation, and self-poisoning by strong adsorbed byproducts. Generally, the adsorption and catalytic properties of Au depend on its particle size and surface chemistry of support which the latter could be controlled by the preparation method and type of catalyst support [8-10]. One of the most important parameters that can change the nature of support surface is the morphology since it could change the coordination number of the surface atoms.
In this work, the co-precipitation method was employed in order to fabricate gold nanocatalysts with different morphologies in the support. The nanocatalysts used in aerobic oxidation of benzyl alcohol, showed unusual high activity in oxidation of benzyl alcohol. Finally, the concentration of defects were probed by optical and structural analysis to explore unusual defect structure in this ZnO support.

2. Experimental
2.1. Preparation and characterization of Au/ZnO

To synthesize gold catalyst with 1 wt % Au loading, firstly, the $Zn(CH_3COO)_2 \cdot 2H_2O$ and PVP were dissolved in 50 ml DI water with a 10:1 ratio (Zn/PVP) and then, an aqueous solution $HAuCl_4.3H_2O$ (5 mL, 0.06 mol/L) was added to this aqueous solution while being stirred at 80

°C. Aqueous sodium carbonate (0.25 mol/L, Aldrich) was added drop wise until a pH of 8.0 was obtained. The material was recovered by filtration and they washed several times with cold and then hot water to ensure removal of Na⁺ and Cl⁻. After drying, initially at room temperature, and then in an oven at 90 °C for 12 h, samples were also annealed in static air at 300 °C and 600 °C for 2 hours.

2.3. Activity measurement of catalyst

The reactions were carried out in a three-necked batch reactor (30 ml) provided with an electronically controlled magnetic stirrer connected to a large reservoir (5000 ml) containing oxygen at 1.5 atm. The reactor was fitted with a reflux condenser, water bath and also thermocouple to control various temperature from 20 °C to 80 °C. The oxygen uptake was followed by a mass flow controller connected to a PC through an A/D board, plotting a flow/time diagram. For the catalytic experiment, a mixture of Na_2CO_3 in toluene solution (0.35 mol/L, 25 ml) and 0.1 g of catalyst was prepared in the three-necked flask. The 0.2 ml of benzyl alcohol was then added into the solution and the resulting mixture was stirred at desired temperatures with a stirring speed of 800 rpm for 3 hours. After the reaction, the catalyst was removed from the reaction mixture by centrifugation. Recoveries were always 97 ± 2% with this procedure. For the identification and analysis of the products and unconverted reactants, a GC (Dani Gas Chromatograph equipped with a flame ionization detector and CP-Sil 8 CB column) was used. For the quantification of the reactant products the calibration method using an external standard was employed.

3. Discussion
3.1. Catalyst characterization

SEM images (Figure 1) show the effect of annealing temperatures on morphology of support.

Figure 1. The FE-SEM images of annealed ZnO nanostructures with various morphologies before doing reaction. A) Nanowies made by annealing at 300 °C. The average diameter is 71 nm. B) Nanoparticles produced by annealing at 600 °C. The average size is 67 nm. The scale bar is 100 nm and Magnification is 40000.

In the sample annealed at 300 °C, ZnO nanowires (ZnO-NW) were formed with an average length of about 2 μm and diameter of 71 nm in average (Figure 1.A). If the annealing temperature increases to 600 °C , ZnO nanopaticles (ZnO-NP) will produce. The mean diameter of ZnO nanoparticles was 67 nm with 45-90 nm size distribution determined by SEM (Figure 1.B).

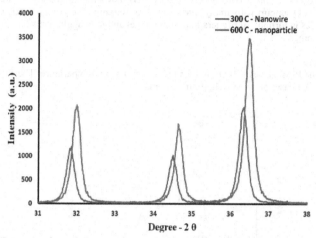

Figure 2. The XRD patterns of annealed ZnO nanostructures with various morphologies. A) Nanoparticles produced by annealing at 600 °C B) Nanowire produced by annealing at 300 °C . Three peaks of two morphologies for identify changing in intensity and shift to right in peaks.

Figure 2 shows X-ray diffraction patterns of ZnO nanowire and nanoparticles. All diffraction peaks can be indexed as wurtzite structure ZnO (JCPDS 36–1451, a=3.24982 Å, c=5.20661 Å). As shown at figure 2, by increasing the annealing temperature and subsequently converting the wire morphology to spherical one, peak positions have been shifted to right and also the intensity of peaks are increased. The origin of this shift can be related lattice strain due to changing morphology and making new defects but it is open question and needs to more investigations.

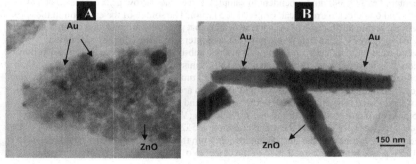

Figure 3. A) The TEM images of annealed ZnO nanostructures after doing reaction. A) Nanoparticles produced by annealing at 600 °C. The average size is 67 nm and the scale bar is 50 nm. B) Nanowies produced by annealing at 300 °C. The average diameter is 71 nm and the scale bar is 20 nm It is indicated that the size of gold nanoparticles is between 4 nm and 10 nm.

TEM image of Au/ZnO samples (Figure 3.A , Figure 3.B) shows that gold particles are highly dispersed on the ZnO support with an average size below 10.0 nm. The growth mechanism of ZnO nanostructures is forming of primers in polymeric gel and then agglomeration to reach a size about 60-70 nm.

3.3. Defect structure

Figure 4 shows the PL spectra of ZnO samples measured at room temperature. The PL was excited by a He – Cd laser at a wavelength of 325 nm.

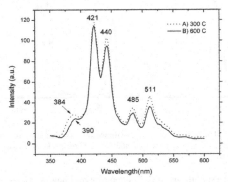

Figure 4. The PL spectra of our samples with a 325 nm excitation wavelength. Our smples were annealed ZnO nanostructures with nanowire and nanoparticle morphologies. Five major peaks are 385, 421, 441, 485 and 511 nm.

The PL spectra exhibits a unique feature relative to other works [11-19], with one UV broad peak centered at 384-390 nm (dependent to sample), three blue narrow peaks centered at 421, 440, 485 and one green narrow peak centered at 511 nm. The source of the emission in the visible-light region attributed to ZnO surface detects that generally are included oxygen vacancy (V_o), zinc vacancy (V_{Zn}), interstitial oxygen (O_i), and interstitial zinc (Zn_i). The emission originated from these defects can be divided to three sub-band included blue, green and yellow emission. It is reported that 410, 440 nm peaks (blue emission) are related to V_{Zn} and Zn_i respectively [12] although there is another peak at 650 nm [25] (in visible range) created by zinc vacancy. The luminescence bands at 421 and 484.6 nm are caused by the transition from the level of the ionized oxygen vacancies to the valence band [20, 21]. It is also believed that 500 and 550 nm emission bands are originated from double ($V_o^{..}$) and single charged ($V_o^{.}$) oxygen vacancies, respectively [13] and the green emission at 532 nm is attributed to the presence of the ionized oxygen vacancy on the surface and results from the recombination of a photogenerated hole with a single ionized charged state of the defect (Vo $^{-}$) in ZnO [22]. The presence of O_i

defect causes the transitions of excited electrons from the valence band to the level of O_i occurs at 565 nm peak (yellow emission) in ZnO nanostructures [14]. The intensity of the green emission (511 nm peak) attributed to oxygen vacancies may indicate difference between two ZnO samples because the intensity of green emission in ZnO sample annealed at 300 °C is stronger than ZnO samples annealed at 600 °C (Figure 4). Based on our knowledge, presence of Z_{ni} (440 nm), V_o (511 nm) defects are definitely in our samples but there is not any content of O_i, V_{Zn} defects and also Vo^-. It seems that surface defects that include Zn and Vo defects and not O_i, V_{Zn} show considerable influence in developing of proper Au nanoclusters for catalytic oxidation.

3.4. Benzyl alcohol oxidation

The oxidation of benzyl alcohol is often used as a model reaction for aromatic alcohol oxidation. Conversion of benzyl alcohol to benzaldehyde was examined by 1 wt% Au nanoparticles loading for both ZnO nanoparticles and nanowires at different reaction temperatures for 3 h. The major product was benzaldehyde but benzyl benzoate and benzoic acid were formed in minor amounts. There is no activity for the samples at 20 °C. The best conversion of benzyl alcohol on Au/ZnO-NW and Au/ZnO-NP catalysts occurred at 60 °C and 80 °C, respectively. In Au/ZnO-NW, the benzaldehyde selectivity was slightly changed from 98.1%, whereas the benzyl alcohol conversion depended on the reaction temperature. A poor conversion of 5.3% was obtained at 20 °C but it reached to 91.4% at 60 °C. In continue, at 80 °C, the alcohol conversion and benzaldehyde yield decreased to 87.7 % and 82.0 %, respectively. A decrease in conversion from 91.4% to 87.7% has been observed at temperature of 80 °C because of reduced dissolved oxygen concentrations. In Au/ZnO-NP, the benzaldehyde selectivity was changed from 92.3 % at 20 °C to 85.1 % at 80 °C whereas the yield reached from 3.5 % to 69.9 % ,respectively. It was conluded that Au/ZnO-NW catalyst was more active than Au/ZnO-NP catalyst especially in higher temperatures. Oxidation of benzyl alcohol on ZnO support has been reported in some works such as Zheng et al. [23] and Choudhary et al. [24] that their reaction was done at 100 °C and 130 °C , respectively. The comparative results were shown in Table 1. One reason to explain high activity of these catalysts can be related to unusual surface defect structure that revealed by PL spectroscopy but more accurate explanation needs to more investigation.

Table 1. Catalytic results for the oxidation of benzyl alcohol with O_2 over ZnO-NW & ZnO-NP supported gold catalysts prepared by the co-precipitation method and comparison with other research work. Our reaction conditions: 0.1 g catalyst, 10 mmol benzyl alcohol, 3 h, 5 mL/min O_2.

Catalyst	Conversion %	Selectivity %	Yield %	Time of reaction (h)	Temperature of reaction (°C)	Ref.
2.5 % on TiO$_2$	74.5	96.4	71.8	6	100	45
2.5 % on ZnO	68.1	81.5	55.5	5	100	42
6.6 % on ZnO	70.5	92.8	65.4	5	130	44

| 1 % on ZnO-NW | 91.4 | 93.7 | 85.6 | 3 | 60 | - |
| 1 % on ZnO-NP | 82.1 | 85.1 | 69.9 | 3 | 80 | - |

4. Conclusion

In summary, Au/ZnO catalysts with two different morphologies was successfully synthesized by using CP method. It was found that the prepared nanocatalysts are highly active in the oxidation of benzyl alcohol to benzaldehyde. An activity of >93% and a selectivity of ~91% was obtained for Au/ZnO-nanowire catalyst at 60 °C whereas Au/ZnO-nanoparticle catalyst exhibited an activity of >85% and selectivity of ~82% at 80 °C. Due to the presence of more low coordinated atoms, ZnO nanowires show better results for trapping and activating the gold atoms for benzyl alcohol oxidation in comparison with Au/ZnO-nanoparticle. PL showed that the concentration of oxygen vacancies in ZnO-nanowire support is more than ZnO-nanoparticle support. The strong interaction between O-vacancy located on ZnO nanowires and Au atoms improves its catalytic behavior in comparison with Au/ZnO nanoparticle catalyst.

5. References

[1] M. Hudlicky, Oxidation in Organic Chemistry, American Chemical Society, Washington, DC, 1990.
[2] M. Haruta, T. Kobayashi, H. Sano, N. Yamada, Chem. Lett., 16, 405 (1987).
[3] A. Abad, P. Concepcion, A. Corma, H. Garcia, Angew. Chem. Int. Ed., 44, 4066 (2005).
[4] H. Miyamura, R. Matsubara, Y. Miyazaki, S. Kobayashi, Angew. Chem. Int. Ed., 46, 4151 (2007).
[5] F.Z. Su, Y.M. Liu, L.C. Wang, Y. Cao, H.Y. He, K.N. Fan, Angew. Chem. Int. Ed., 47, 334 (2008).
[6] J.M. Campelo, T.D. Conesa, M.J. Gracia, M.J. Jurado, R. Luque, J.M. Marinas, A.A. Romero, Green Chem. , 10, 853 (2008).
[7] J.J. Zhu, J.L. Figueiredo, J.L. Faria, Catal. Comm. , 9, 2395 (2008).
[8] L.C. Wang, Q. Liu, X.S. Huang, Y.M. Liu, Y. Cao, K.N. Fan, Appl. Catal. B: Environ., 88, 204 (2009).
[9] M. Huruta, M. Date, Appl. Catal. A: Gen., 222, 427 (2001).
[10] I. Dobrosz-Gomez, I. Kocemba, J.M. Rynkowski, Appl. Catal. B: Environ. , 88, 83 (2009).
[11] M. Ramani, S. Ponnusamy, C. Muthamizhchelvan, Opti. Mater., 34, 817 (2012).
[12] E. Rauwel, A. Galeckas, P. Rauwel, M.F. Sunding, H. Fjellvag, J. Phys. Chem. C, 115, 25227 (2011).
[13] Y. Lai, M. Meng, Y. Yu, X. Wang, T. Ding, Appl. Cata. B: Environmental, 105, 335 (2011).
[14] W.D. Yu, X.M. Li, X.D. Gao, P.S. Qiu, W.X. Cheng, A.L. Ding, Appl. Phys. A 79, 453 (2004).
[15] Y.W. Wang, L.D. Zhang, G.Z. Wang, X.S. Peng, Z.Q. Chu, C.H. Liang, J. Crys. Grow. , 234, 171 (2002).

[16] J. Yang, X. Liu, L. Yang, Y. Wang, Y. Zhang, J. Lang, M. Gao, M. Wei, J. Alloy. Comp. , 485, 734 (2008).

[17] J.D. Ye, S.L. Gu, F. Qin, S.M. Zhu, S.M. Liu, X. Zhou, W. Liu, L.Q. Hu, R. Zhang, Y. Shi, Y.D. Zheng, Appl. Phys. A: Mater. Sci. Process 81, 759 (2005).

[18] Y. Zheng, C. Chen, Y. Zhan, X. Lin, Q. Zheng, K. Wei, J. Zhu, Y. Zhu, Inorg. Chem., 46, 6675 (2007).

[19] J. Zhang, L. Sun, J. Yin, H. Su, C. Liao, C. Yan, Chem. Mater., 14, 4172 (2002).

[20] D. H. Zhang, Q. P. Wang, and Z. Y. Xue, Appl. Surf. Sci. 207, 20 (2003).

[21] G. H. Du, F. Xu, Z. Y. Yuan and G. Van Tendeloo , Appl. Phys. Lett. 88, 243101 (2006).

[22] K. Vanheusden, W. L. Warren, C. H. Seager, D. R. Tallant, J. A. Voigt, and B. E. Gnade, J. Appl. Phys. 79, 7983 (1996).

[23] N. Zheng , Galen D. Stucky , Chem. Commun., 6, 3862(2007).

[24] V. R. Choudhary, A. Dhar, P. Jana, R. Jha and B. S. Uphade , Green Chem. , 7, 768 (2005).

[25] D. C. Iza, D. Muñoz-Rojas, Q. Jia, B. Swartzentruber, J. L MacManus-Driscoll, Nano. Res. Lett. 2012, 7, 655.

Mater. Res. Soc. Symp. Proc. Vol. 1675 © 2014 Materials Research Society
DOI: 10.1557/opl.2014.903

Synthesis of $Zn_xMg_{1-x}O$ Nanocrystals and the Assessment of their Antimicrobial Activity against *Escherichia Coli*

Yarilyn Cedeño-Mattei[1,2], Rosa Concepción-Abreu[3], and Oscar Perales-Pérez[1,2]

[1]Department of Engineering Science and Materials, University of Puerto Rico, Mayaguez, PR 00681-9000, U.S.A.
[2]Department of Chemistry, University of Puerto Rico, Mayaguez, PR 00681-9000, U.S.A.
[3]Department of Biology, Chemistry, and Environmental Sciences, Interamerican University of Puerto Rico, San Germán, PR 00683, U.S.A.

ABSTRACT

The present work focuses on the synthesis and evaluation of the antimicrobial activity of $Zn_xMg_{1-x}O$ solid solutions. $Zn_xMg_{1-x}O$ solid solutions were synthesized through the thermal decomposition of ZnMg-precursor synthesized in aqueous and ethanol solutions via a two-steps process. The antimicrobial activity of $Zn_xMg_{1-x}O$ solid solution against *E. coli* was evaluated using the spread plate method in presence of $Zn_xMg_{1-x}O$ powder of different contents of Zn species, 'x'. The powder concentrations evaluated were 500, 1000, and 1500 ppm. $Zn_{0.10}Mg_{0.90}O$ powders exhibited a bacterial growth inhibition between 38% and 100% when the powder concentration increased from 500 up to 1500 ppm, respectively. A decreasing trend was observed for x = 0.30 and above; the corresponding bacterial growth inhibition was 12%, 6%, and 5% when the particles concentration was, respectively, 500, 1000, and 1500 ppm. X-Ray diffraction analyses suggested the incorporation of Zn ions into the MgO lattice for 'x' values below 0.10, enhancing the antimicrobial activity; the formation of two isolated oxide phases observed at larger 'x' values (e.g. x = 0.30 and x = 0.50 Zn), could explain the detected inhibition of the corresponding antimicrobial activity.

INTRODUCTION

Food packaging is indispensable to preserve the quality and safety of the food from the time of manufacturing to the final use by the consumer. It is indispensable to evaluate new materials with enhanced antimicrobial capacity that could be then dispersed in biodegradable polymers for food packaging applications. ZnO and MgO have been suggested as antimicrobial agents among others [1–3]. It would be expected that synthesizing a unique nanostructure containing Mg and Zn species as $Zn_xMg_{1-x}O$ solid solutions, should be conducive to a synergistic effect and enable the tailoring in the corresponding antimicrobial activity to a broader kind of microorganisms. Taking into account the above considerations, the present work addresses the development of a novel protocol to synthesize $Zn_xMg_{1-x}O$ nanocrystals, their structural and morphological characterization, and the composition-based assessment of the corresponding antimicrobial activity in presence of *E. coli*.

EXPERIMENTAL

Materials
Suitable amounts of $Mg(NO_3)_2$ (ACS, 98-102%, Alfa Aesar), $Zn(OOCCH_3)_2 \cdot 2H_2O$ (ACS, 98-101%, Alfa Aesar), Na_2CO_3 ($\geq$ 99%, Sigma-Aldrich), and NaOH (pellets, 98%, Alfa Aesar) were used for the synthesis of the Mg-Zn carbonate hydroxide precursor. All regents were used without any further purification.

Synthesis of Zn-Mg Hydroxide Carbonate Precursor and $Zn_xMg_{1-x}O$ Solid Solution
The synthesis method employed in the formation of Zn-Mg precursor corresponds to a modification of the Y. Zhao *et al* method [4]. The method involves the precursor formation in separate nucleation and aging steps. The samples were synthesized in aqueous phase and using an ethanol/water mixture 80/20 v/v as solvent. Solution A consisted of 100 mL of suitable amounts of $Mg(NO_3)_2$ and $Zn(OOCCH_3)_2 \cdot 2H_2O$ according to the $Zn_xMg_{1-x}O$ stoichiometry (x = 0.00-0.50) while solution B contained 100 mL of stoichiometric amounts of Na_2CO_3 and NaOH. The nucleation step involved the mixing of solutions A and B for 2 minutes at 11000 rpm. The resulting slurry was aged for one hour at 100°C. The solid precursor was washed three times either with deionized water or ethanol/water mixture, depending on the synthesis medium and dried at 50 °C for 24 hours. The Mg-Zn carbonate hydroxide precursor was thermally treated in order to promote the formation of $Zn_xMg_{1-x}O$. The thermal treatment conditions were set based on previous experiments: 600 °C for 1 hour in air using a 10 °C/min heating ramp.

Nanomaterials Characterization
A Siemens D500 X-Ray Diffractometer with Cu K_α radiation was used to analyze the crystallinity of precursors and solid solutions, in addition to study either the incorporation of Zn into the MgO structure or the formation of secondary phases. The average crystallite size was calculated using the Scherrer's equation [5]. The morphology, composition, and nanometric nature of the materials were examined using a JEM-ARM200cF Transmission Electron Microscope.

Antimicrobial Activity
In general, the antimicrobial activity of $Zn_xMg_{1-x}O$ against *Escherichia Coli* (ATCC 35218) was conducted using the Spread Plate Method [6]. A cell culture Erlenmeyer flask containing 10 mL Tryptic soy broth was contacted with specific $Zn_xMg_{1-x}O$ powder concentrations (500, 1000, and 1500 ppm). The test units were placed in an incubator at 37°C and 250 rpm. After 24 hours of incubation period, the samples were submitted to serial dilutions until reach a dilution factor of 10^7. A small amount of the bacterial suspension (0.1 mL) was spreaded over the Mueller Hinton agar surface and incubated at 37 °C for 24 hours to permit the growth of colonies. The colonies were subsequently quantified and multiplied for the corresponding dilution factor in order to calculate the bacterial growth inhibition.

DISCUSSION

XRD Analyses

Figure 1-a shows the XRD patterns of Zn-Mg carbonate hydroxide precursors synthesized in aqueous phase at different 'x' values. Only those peaks corresponding to hydromagnesite (magnesium carbonate hydroxide) were observed for those samples up to 0.10 in Zn content. Broader and noisier diffraction peaks were observed in the patterns corresponding to the sample synthesized at 'x' of 0.30 and 0.50, suggesting the formation of a less crystalline solid. Diffraction peaks around 34 and 58° became prominent at 'x' of 0.30 and 0.50. Those peaks are attributed to ZnO – wurtzite phase. In turn, figure 1-b shows the XRD patterns corresponding to the same precursor but now synthesized in an ethanol/water mixture 80/20 v/v. Ethanol was used in order to take advantage of its dehydrating nature ad accelerate the formation of the final solid. Generally speaking, less crystalline precursors were formed in comparison to those synthesized in aqueous phase for a similar 'x' value.

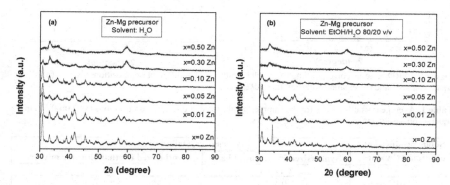

Figure 1. XRD patterns of Zn-Mg hydroxide carbonate precursors synthesized at different compositions, 'x', in: (a) aqueous phase and (b) ethanol/water mixture 80/20 v/v.

Figure 2-a shows the XRD patterns for the samples formed after thermal treatment at 600 °C for 1 hour in air of the Zn-Mg precursors synthesized in water. The crystallographic planes (111), (200), (220), (311), and (222), corresponding to cubic MgO-periclase, became evident for 'x' below 0.10. The XRD analysis of the sample produced at x = 0.10 revealed the incipient presence of ZnO-wurtzite. The ZnO peaks are more noticeable in those samples synthesized at x = 0.30 and 0.50, and co-exist with those of periclase. The XRD patterns for the Zn-Mg products that evolved from the precursor synthesized in ethanol/water mixture are shown in Figure 2-b. On a general basis, the patterns suggest the same trends as in the samples synthesized in aqueous phase. The main difference arises from peaks broadening. The samples synthesized in ethanol/water mixture exhibit sharper peaks, which result in bigger average crystallite sizes as shown in table 1. The average crystallite sizes of the Zn-Mg oxides synthesized in aqueous phase ranged between 8-13 nm, which are smaller than the 15-20 nm corresponding to those samples produced in the ethanol/water mixture. The rationale behind these differences can be explained as follows: solids generated from highly crystalline precursors (like those formed in water) led to smaller crystal sizes; highly ordered structures demands more energy to disturb the original structure and cause the atomic rearrangement associated to the formation of the final oxides, and subsequent crystal growth. Otherwise, disordered structures (like the ones generated in the ethanol/water mixture) rearrange easier and employ the excess of energy to promote crystal growth [7].

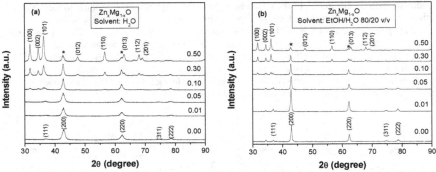

Figure 2. XRD patterns corresponding to samples synthesized at different compositions, 'x', in: (a) aqueous phase, and (b) ethanol/water mixture 80/20 v/v.

Table 1. Average crystallite sizes for the samples synthesized at different compositions (x=0.00-0.50) in aqueous phase and ethanol/water mixture 80/20 v/v.

Solids produced at different 'x' values	Average crystallite size (nm)	
	Synthesized in aqueous phase	Synthesized in ethanol/water mixture
0.00	8.3± 0.4	15 ± 1
0.01	8.6 ± 0.3	19 ± 2
0.05	8.0 ± 0.2	18 ± 2
0.10	10 ± 3	20 ± 3
0.30	13.5 ± 0.7	19 ± 3
0.50	11 ± 3	15 ± 1

Transmission Electron Microscopy Analyses
Figure 3 shows the TEM and HRTEM images, and size distribution histograms of the samples synthesized in aqueous phase at 'x' = 0.05 (a, b, and c) and 0.30 (d, e, and f). The nanometric nature of the particles and their high crystallinity were evidenced. Moreover, the particle size estimated from TEM observations are slightly bigger, but in good agreement, than the average crystallite sizes calculated using the Scherrer's equation that suggests the formation of single crystals. The sizes calculated from TEM observations are: 12 ± 3 and 18 ± 5 for Zn content 'x' = 0.05 and 0.30, respectively.

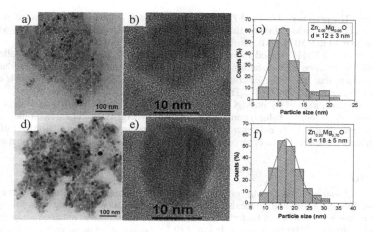

Figure 3. TEM, HRTEM images, and size distribution histograms of samples synthesized in aqueous phase at different 'x' values: 0.05 (a, b, and c) and 0.30 (d, e, and f).

Assessment of Antimicrobial Activity of Zn$_x$Mg$_{1-x}$O Solid Solution

The *E. coli* growth inhibition in presence of Zn-MgO nanocrystalline powders synthesized in aqueous phase is shown in Figure 4.

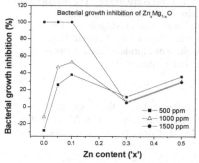

Figure 4. Bacterial growth inhibition percentage for *E. coli* in presence of: pure MgO and Zn-Mg oxides synthesized at different 'x' values in aqueous phase. The bacteria were contacted with 500, 1000, and 1500 ppm of powdered samples in each case.

Although 1500 ppm of pure MgO nanocrystals allowed a complete inhibition of bacterial growth, a concentration of 500 ppm promoted the bacterial growth, instead. In turn, Zn$_x$Mg$_{1-x}$O solid solutions synthesized at 'x' ≤ 0.10 inhibited bacterial growth even at 500 ppm of the solids concentration (26% and 38% of inhibition for 'x' of 0.05 and 0.10, respectively). This synergistic effect on the inhibition of bacterial growth can be explained in terms of the actual formation of the Zn-Mg solid solutions, where Zn species could have replaced Mg sites in the MgO host lattice. On the contrary, the inhibition went down to 5% even for a particle concentration as high as 1500 ppm for Zn contents over 'x' = 0.30. This later effect would be due to the formation of two-isolated oxide phases (Zn$_x$Mg$_{1-x}$O and ZnO) instead of the solid solution, as evidenced by

XRD analyses. The presence of isolated ZnO could have acted as a nutrient source instead of a strong bactericidal compound. In turn, the observed increase in bacterial growth inhibition when 'x' varied from 0.30 to 0.50 could be attributed to the smaller size of the nanoparticle synthesized at 'x' = 0.50 (11 nm and 13.5nm, respectively, for 'x' 0.50 and 0.30). The smaller the nanoparticle, the more active it becomes and hence, more electrons are available to produce the redox reactions that lead to the generation of reactive oxygen species (ROS), responsible of the antimicrobial activity of ZnO [8]. The antimicrobial activity of MgO has been mainly attributed to the generation of O_2^- on MgO surface [9]. Therefore, the formation of a $Zn_xMg_{1-x}O$ solid solution (for suitable 'x' values) could have induced a synergistic effect based on the probable co-generation of ROS and superoxide bactericidal species. The synergistic effect of the Zn concentration in the solids was observed up to 'x' = 0.10 for particles concentrations below 1500 ppm.

CONCLUSIONS

$Zn_xMg_{1-x}O$ solid solutions and Zn-Mg composite oxides were successfully synthesized in the 'x' range 0.00-0.50 using a two-step synthesis method based on nucleation and aging steps followed by thermal treatment at 600 °C. A bacterial growth inhibition of 100% was obtained for high particles concentration (1500 ppm) and 'x' below 0.10 for $Zn_xMg_{1-x}O$ solid solutions synthesized in aqueous phase. $Zn_xMg_{1-x}O$ solid solutions exhibit an enhanced inhibitor behavior when compared to pure MgO at low nanoparticle concentration (e.g. 500 and 1000 ppm).

ACKNOWLEDGMENTS

This project is supported by the Agriculture and Food Research Initiative Competitive Grant No. 2012-67012-19806 from the USDA-National Institute of Food and Agriculture. The authors also thank to the USDA-NIFA Center for Education and Training in Agriculture and Related Sciences (CETARS), Competitive Grant No. 2011-38422-30835. The TEM facility at FSU is funded and supported by the Florida State University Research Foundation, National High Magnetic Field Laboratory (NSF-DMR-0654118) and the State of Florida. The contribution from Dr. Félix R. Román, Chemistry Department at UPRM, is greatly appreciated.

REFERENCES

1. S. Ravikumar, R. Gokulakrishnan, and P. Boomi, Asian Pac. J. Trop. Dis., **2**, 85 (2012).
2. O. Yamamoto, T. Ohira, K. Alvarez, and M. Fukuda, Mater. Sci. Eng. B, **173**, 208 (2010).
3. R. Brayner, R. Ferrari-Iliou, N. Brivois, S. Djediat, M.F. Benedetti, and F. Fiévet, Nano Lett., **6**, 866 (2006).
4. Y. Zhao, F. Li, R. Zhang, D.G. Evans, and X. Duan, Chem. Mater., **14**, 4286 (2002).
5. B. D. Cullity, in Elements of X-Ray Diffraction, edited by Morris Cohen (Addison Wesley, MA, 1972), p. 102.
6. M.T. Madigan, J.M. Martinko, P.V. Dunlap, and D.P. Clark, *Brock Biology of Microorganisms*, 13th ed. (Benjamin Cummings, San Francisco, CA, 2008), p. 129.
7. Y. Cedeño-Mattei, M. Reyes, O. Perales-Perez, and F. Román Mater. Res. Soc. Symp. Proc. 2013, 1547, 135. doi:10.1557/opl.2013.638.
8. A. Akbar and A. K. Anal, Food Control. **38**, 88 (2014).
9. J. Sawai, E. Kawada, F. Kanou, H. Igarashi, A. Hashimoto, T. Kokugan, and M. Shimizu, J. Chem. Eng. Jpn., **29**, 627 (1996).

Multiferroics, Magnetism, and Magnetic Materials

Mater. Res. Soc. Symp. Proc. Vol. 1675 © 2014 Materials Research Society
DOI: 10.1557/opl.2014.800

Persistent Photoconductivity in Bulk Strontium Titanate

Matthew D. McCluskey,[1] Caleb D. Corolewski,[1] Violet M. Poole,[1] and Marianne C. Tarun[1]

[1]Washington State University, Pullman, WA 99164-2814, U.S.A.

ABSTRACT

Strontium titanate ($SrTiO_3$) has novel properties, including a large temperature-dependent dielectric constant, and can be doped to make it metallic or even superconducting. The origin of conductivity observed at the $SrTiO_3/LaAlO_3$ interface is a topic of intense debate. In the present work, bulk single crystal $SrTiO_3$ samples were heated at 1200°C, with the goal of producing cation vacancies. These thermally treated samples exhibited persistent photoconductivity (PPC) at room temperature. Upon exposure to sub-band-gap light (>2.9 eV), the free-electron density increases by over two orders of magnitude. This enhanced conductivity persists in the dark, at room temperature, for several days with essentially no decay. Light excites an electron from the vacancy to the conduction band, where it remains, due to a large recapture barrier. These observations highlight the importance of defects in determining the electrical properties of oxides and may point toward novel applications.

INTRODUCTION

Oxide semiconductors and insulators are used in a broad range of devices and are actively investigated for applications in energy production and storage, optoelectronics, and transparent electronics. The range of properties exhibited by oxide materials provides numerous possibilities for novel electronic devices [1]. Oxides show physical phenomena including ferroelectricity [2], high-temperature superconductivity [3], and colossal magnetoresistance [4]. These materials have a complex phase behavior that is often highly sensitive to temperature, doping, or the application of stress. Heterointerfaces may produce unexpected properties. Famously, the interface between the insulating oxides strontium titanate ($SrTiO_3$) and lanthanum aluminate ($LaAlO_3$) results in a highly conductive layer [5,6]. The physical mechanisms behind this conductivity are a subject of intense debate.

Our initial studies investigated hydrogen in $SrTiO_3$. In as-grown $SrTiO_3$ bulk samples, hydrogen forms an O-H bond with a host oxygen atom, resulting in a bond-stretching vibration near 3500 cm^{-1} [7]. In samples annealed in hydrogen, Tarun and McCluskey [8] found this peak (H_I) along with new peaks at 3355 and 3384 cm^{-1} (H_{II}). The corresponding O-D peaks showed the expected isotope shift to lower vibrational frequency. Samples annealed in a hydrogen-deuterium mixture showed new H_{II} peaks. These peaks corresponded to HD centers, showing that the H_{II} center consists of two hydrogen atoms.

The microscopic structures of these hydrogen complexes are not known. Theoretical studies have suggested that the H_I line arises from an isolated hydrogen interstitial [9] while H_{II} represents a pair of hydrogen interstitials that are attracted to each other via lattice strain [10]. Tarun and McCluskey [11] attributed H_{II} to a strontium vacancy that is fully passivated by two hydrogen atoms. However, the experimental results cannot exclude the possibility that H_{II} is a titanium vacancy decorated by two hydrogen atoms [12,13], or two neighboring hydrogen interstitials [10]. Furthermore, one must consider the possibility that the H_I configuration is a defect-hydrogen complex rather than simply isolated hydrogen [12]. This is important because identifying the hydrogen interstitial is required to test the universal hydrogen model.

To determine whether the H_{II} complex involves a vacancy, we annealed samples in a sealed ampoule, in the presence of SrO or TiO_2 powder. The goal was to observe a correlation between the H_{II} peaks and Sr or Ti vacancies. In fact, annealing in different ambient conditions did not affect the intensity of the H_{II} peaks. However, an unexpected discovery was made. Annealed samples showed persistent photoconductivity (PPC), defined as an increase in electrical conductivity after exposure to light, which persists after the light is turned off.

Room-temperature PPC has been observed previously in semiconductor [14,15] and superconductor [16] materials. These samples generally show an increase in conductivity that is less than an order of magnitude. The enhanced conductivity decays according to a stretched exponential or similar function, with a time constant of several hours. Large PPC (greater than an order of magnitude) has been observed due to DX centers in GaAs and other semiconductors [17,18], but only at low temperatures ($T < 100$ K). Our observation of PPC in $SrTiO_3$ crystals is unique because it is a large effect that is observed at room temperature.

EXPERIMENT

Verneuil-grown $SrTiO_3$ single crystals were purchased from MTI Corporation and MaTeck GmbH. Samples were sealed in an evacuated fused silica ampoule, along with SrO powder, TiO_2 powder, or Ar. Annealing was conducted in a three-zone horizontal tube furnace at 1200°C for 1 hr. After annealing, the ampoule was taken out of the furnace and allowed to cool to room temperature in the dark. We found that samples annealed with TiO_2 powder or Ar were highly n-type, perhaps related to the creation of oxygen vacancies or other decomposition processes. Samples annealed in SrO powder were insulating. All annealed samples showed PPC.

Free-carrier absorption spectra were obtained using a Bomem DA8 vacuum Fourier transform infrared spectrometer with a globar light source, KBr beamsplitter, and liquid-nitrogen-cooled InSb detector. Hall-effect measurements (MMR Technologies) were performed in the van der Pauw geometry. Melted indium was used to make electrical contacts. To probe the wavelength dependence of the photoconductivity, we used the 450-W Xe lamp of a JY-Horiba FluoroLog-3 spectrofluorometer as a source of monochromatic illumination. Optical transmission spectra were obtained with a Perkins-Elmer UV/visible/IR spectrometer.

DISCUSSION

When exposed to sub-band-gap light, the free-electron concentration increases by over two orders of magnitude (Fig. 1) [19]. After the light is turned off, the enhanced conductivity persists for days with negligible decay at room temperature. The same effect is observed using IR spectroscopy, in which free-carrier absorption increases significantly after illumination (Fig. 2). We attribute the large PPC to the excitation of an electron from a vacancy defect to the conduction band, with an extremely low recapture rate.

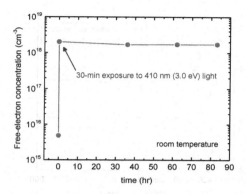

Figure 1. Persistent photoconductivity in strontium titanate (MTI, 1 mm thick). The free-electron concentration increases by over 2 orders of magnitude after exposure to light. The enhanced conductivity persists in the dark.

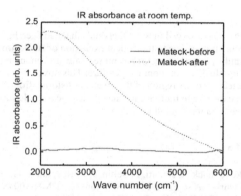

Figure 2. IR absorption spectrum of an $SrTiO_3$ crystal (MaTeck GmbH, 0.5 mm thick), before and after exposure to light.

Optical transmission spectra, referenced to an as-received sample, show absorption peaks at 425 nm (2.9 eV) and 510 nm (2.4 eV). The 2.9 eV peak has been attributed to Fe^{4+} impurities [20] and is often observed in annealed samples [21]. The 2.4 eV peak has been observed in thermally treated samples but has not been identified conclusively [22]. Since this peak increases upon illumination, it may be a signature of the defect responsible for PPC.

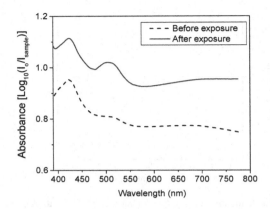

Figure 3. Optical absorption spectrum of an annealed SrTiO₃ crystal before and after exposure to light

CONCLUSIONS

Room temperature PPC was observed in $SrTiO_3$ crystals that had been heated to 1200°C. The large enhancement in conductivity is due to the optical excitation of an electron from a defect level to the conduction band. A large barrier for recapture results in conductivity that persists for days (or longer) with negligible decay at room temperature. This novel effect could enable new technologies such as 3D electronics. A region of the crystal transforms from an insulator to a conductor simply through exposure to light. Hence, an "optical pen" could be used to sketch a circuit on the crystal surface and throughout the interior.

ACKNOWLEDGMENTS

The authors would like to thank A. Janotti, F. Selim, J. Varley, and C.G. Van de Walle for helpful discussions. This work was supported by NSF Grant No. DMR-1004804, with partial student support provided by DOE Grant No. DE-FG02-07ER46386.

REFERENCES

1. R. Ramesh and D.G. Schlom, *MRS Bulletin* **33** (No. 11), 1006 (2008).
2. M. Dawber, K. M. Rabe and J. F. Scott, *Reviews of Modern Physics* **77**, 1083 (2005).
3. J.G. Bednorz and K.A. Müller, *Z. Phys. B: Condens. Matter* **64**, 189 (1986).
4. A.P. Ramirez, *J. Phys.: Condens. Matter* **9**, 8171 (1997).
5. A. Ohtomo and H.Y. Hwang, *Nature* **427,** 423 (2004).

6. S.A. Chambers, M.H. Engelhard, V. Shutthanandan, Z. Zhu, T.C. Droubay, L. Qiao, P.V. Sushko, T. Feng, H.D. Lee, T. Gustafsson, E. Garfunkel, A.B. Shah, J.-M. Zuo, and Q.M. Ramasse, *Surf. Sci. Rep.* **65**, 317 (2010).
7. D. Houde, Y. Lépine, C. Pépin, S. Jandl and JL Brebner, *Phys. Rev. B* **35**, 4948 (1987).
8. M.C. Tarun and M.D. McCluskey, *J. Appl. Phys.* **109**, 063706 (2011).
9. L. Villamagua, R. Barreto, L.M. Prócel, and A. Stashans, *Phys. Scr.* **75**, 374 (2007).
10. N. Bork, N. Bonanos, J. Rossmeisl, and T. Vegge, *Phys. Chem. Chem. Phys.* **13**, 15256 (2011).
11. M.C. Tarun and M.D. McCluskey, *J. Appl. Phys.* **109**, 063706 (2011).
12. J.T. Thienprasert, I. Fongkaew, D.J. Singh, M.-H. Du, and S. Limpijumnong, *Phys. Rev. B* **85**, 125205 (2012).
13. J. Varley, A. Janotti, and C.G. Van de Walle, *Phys. Rev. B* **89**, 075202 (2014).
14. M.T. Hirsch, J.A. Wolk, W. Walukiewicz, and E. E. Haller, *Appl. Phys. Lett.* **71**, 1098 (1997).
15. J.Z. Li, J.Y. Lin, H.X. Jiang, J.F. Geisz, and S.R. Kurtz, *Appl. Phys. Lett.* **75**, 1899 (1999).
16. A. Gilabert, A. Hoffmann, M.G. Medici, and I.K. Schuller, *Journal of Superconductivity and Novel Magnetism* **13**, 1 (2000).
17. D.V. Lang and R.A. Logan, *Phys. Rev. Lett.* **39**, 635 (1977).
18. P.M. Mooney, *J. Appl. Phys.* **67**, R1 (1990).
19. M.C. Tarun, F.A. Selim, and M.D. McCluskey, *Phys. Rev. Lett.* **111**, 187403 (2013).
20. B.W. Faughnan, *Phys. Rev. B* **4**, 3623 (1970).
21. B. Jalan, R. Engel-Herbert, T.E. Mates, and S. Stemmer, *Appl. Phys. Lett.* **93**, 052907 (2008).
22. R.J. Wild, E.M. Rockar, and J.C. Smith, *Phys. Rev. B* **8**, 3828 (1973).

Mater. Res. Soc. Symp. Proc. Vol. 1675 © 2014 Materials Research Society
DOI: 10.1557/opl.2014.786

Epitaxial Growth of Ferroelectric Pb(Zr,Ti)O₃ Layers on GaAs

Benjamin Meunier[1], Lamis Louahadj[2], David Le Bourdais[3], Ludovic Largeau[4], Guillaume Agnus[3], Philippe Lecoeur[3], Valérie Pillard[3], Lucie Mazet[1], Romain Bachelet[1], Philippe Regreny[1], Claude Botella[1], Geneviève Grenet[1], David Albertini[5], Catherine Dubourdieu[1], Brice Gautier[5] and Guillaume Saint-Girons[1]

[1]Ecole Centrale de Lyon, INL-CNRS, Ecully, France
[2]RIBER SA, Bezons, France
[3]Université Paris-Sud, Institut d'Electronique Fondamentale, Orsay, France
[4]LPN-CNRS, Marcoussis, France
[5]INSA, INL-CNRS, Villeurbanne, France

ABSTRACT

Ferroelectric epitaxial Pb(Zr,Ti)O₃ (PZT) layers were grown by pulsed laser deposition on SrTiO₃/GaAs templates fabricated by molecular beam epitaxy. The templates present an excellent structural quality and the SrTiO₃/GaAs is abrupt at the atomic scale thanks to surface Ti pre-treatment. The PZT layers contain a- and c- domains, as shown by X-Ray diffraction analyses. Piezoforce microscopy experiments and macroscopic electrical characterizations indicate that PZT is ferroelectric. A relative dielectric permittivity of 164 is extracted from these measurements.

INTRODUCTION

Functional oxides with perovskite structure present a variety of physical properties (ferroelectric, piezoelectric, ferromagnetism,…) that make them very attractive for applications in the micro-optoelectronic field. Combining such oxides in their crystalline form with silicon or III-V semiconductors would allow integrating these functionalities on platforms compatible with industrial applications, and would more generally provide solutions to define novel functionalities or devices based on the combination of the physical properties of oxides and semiconductors (low V_T transistors, MEMS, non-volatile memories,…).

Starting in the late 1990's, it has been shown that MBE, combined with convenient interface engineering strategy, could be used for the epitaxial growth of SrTiO₃ (STO) on Si[1,2] and GaAs[3]. This has been followed closely by two demonstrations of integration of piezoelectric Pb(Zr,Ti)O₃ (PZT) on STO/Si[4]. Demonstrating the integration of ferroelectric or piezoelectric thin layers on semiconductor substrates still remains a challenge, since the chemical reactivity of the semiconductor with oxygen imposes narrow growth windows for the oxide that reduces the room of manoeuvre for optimizing the growth process, and since macroscopic ferroelectric measurement are made difficult in a metal-oxide-semiconductor configuration due to the strong contribution of the semiconductor to the capacitance of the heterostrucure. Within the last two years, piezoelectric lead magnesium niobate-lead titanate (PMN-PT)[5] and PZT[6,7] layers have been integrated on Si using STO templates, and ferroelectricity has been evidenced in BTO layers integrated on Si[8,9] and GaAs[10,11] .

Here, we report on the first demonstration of integration by epitaxy of PZT thin films on GaAs using STO/GaAs templates and a conducting $La_{0.67}Sr_{0.33}MnO_3$ (LSMO) bottom electrode. PLD is used for PZT and LSMO growth, while the STO/GaAs buffers are grown by MBE. Piezoforce microscopy (PFM) as well as macroscopic electrical measurements evidence the ferroelectric behaviour of the PZT films.

EXPERIMENT

The sample considered in the following consists of a 100 nm thick $Pb(Zr_{0.52}Ti_{0.48})O_3$ (PZT) layer deposited on top of a 30 nm thick $(La_{0.67}Sr_{0.33})MnO_3$ (LSMO) electrode. The role of this bottom electrode is explained below. The PZT/LSMO heterostructure was deposited by PLD on a 8 nm thick STO buffer grown by MBE on a GaAs substrate. The epiready GaAs substrate was first introduced in a III-V dedicated MBE chamber and the surface oxide was desorbed by annealing the sample under arsenic (As). A 1 μm thick P-doped GaAs buffer was then grown and capped by a thin amorphous As layer deposited at room temperature. The sample was then exposed to air for transfer into an oxide-dedicated MBE chamber and subsequent STO growth. The protective amorphous As layer was desorbed by annealing the sample under ultra-high vacuum (UHV) at 400°C, leading to the formation of a clean, As-rich, 2×4 reconstructed GaAs surface. Prior to STO deposition, a Ti-based GaAs surface treatment procedure similar to that described in Ref. 3 and 12 was applied : 1/2 monolayer (ML) of Ti were deposited at 400°C leading to the formation of a 4×2 reconstruction of the GaAs:Ti surface. The sample temperature was then ramped down to 250°C, and 6 ML of STO were deposited under an oxygen partial pressure of 5×10^{-8} Torr, leading to the formation of a partially amorphous oxide layer. The sample was then annealed under UHV at 450°C during 20 min, leading to complete recrystallization of the oxide film (Fig. 1(a)). The rest of the STO layer was then grown at this temperature under an oxygen partial pressure of 2×10^{-6} Torr. The STO/GaAs templates were examined by atomic force microscopy (AFM), X-Ray diffraction (XRD) and transmission electron microscopy (TEM) and X-Ray photoelectron spectroscopy (XPS).

LSMO and PZT were grown on such templates by pulsed laser deposition with a KrF laser working at 248 nm with a fluence of 3 J/cm². A homogenizer was used to ensure a top hat profile of the laser beam on the target. The full stack was grown under 120 mTorr of oxygen at 600°C. At the end of the deposition process, the chamber was filled with oxygen up to 300 Torr for 10 min before the cooling down step.

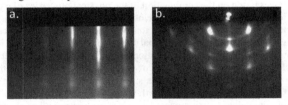

Fig. 1: RHEED patterns are recorded after STO layer growth on a pre-treated GaAs substrate (a) and on a "as-deposited" GaAs substrate (b).

DISCUSSION

Characterization of epitaxial growth STO layer

Fig. 1(a) displays the reflection high energy electron diffraction (RHEED) pattern recorded after the growth of the STO layer on a pre-treated GaAs substrate. It presents well-contrasted streaky lines indicating that the STO surface is flat and single-crystalline. Fig. 1(b) shows the RHEED pattern recorded after the growth of STO on pure GaAs substrate (without Ti pre-treatment of the surface). In this case, Debye-Sherrer rings are detected indicating the formation of a polycrystalline compound.

XPS analyses were carried out using monochromatic AlKα source (1486.6 eV) to study the influence of the Ti pre-treatment on the STO/GaAs interface chemistry. To this end, 7 monolayers (ML) (~2.8 nm) of STO were deposited on GaAs, with and without Ti-pre-treatment. With such low STO thickness, XPS signal arising from the GaAs substrate can be detected. Shirley background and Pseudo-Voigt functions were used for the resolution of the XPS core levels into components. XPS spectra of the Sr3d, As3d, Ti3p and Ga3d core-levels recorded (using maximum intensity normalisation) are displayed in Fig. 2. Fig.2(a),(b) and (c) correspond to the sample for which a Ti-based GaAs surface treatment was used Fig.2(d),(e) and (f) the sample without pre-treatment. Doublets Ga3d$_{3/2}$, Ga3d$_{5/2}$ and As3d$_{3/2}$, As3d$_{5/2}$ are unresolved.

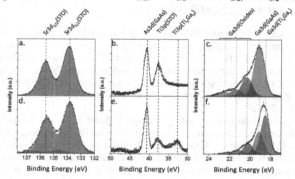

Fig. 2: Sr3d, As3d, Ti3p and Ga3d core levels for 7 ML of STO on GaAs substrate (a), (b) and (c) after ½ ML of Ti pre-deposit; (d), (e) and (f) without Ti pre-deposition.

In Fig 2 a, the Sr3d$_{5/2}$ component at 132.1eV and in Fig 2 b, the Ti3p component at 37.8eV are characteristic of bulk STO. In Fig.2(b), As3d component at 40.8eV peak and in Fig.2(c), the main peak Ga3d at 19.13eV are identified as due to bulk GaAs.[13] The other components in Fig 2 c towards higher binding energy correspond to gallium oxides. They represent 38.5% of the total Ga3d core level area. The major oxide contribution shifted by +1.2 eV from the GaAs component is attributed to Ga$_2$O$_3$. These observations allow for concluding that the GaAs substrate is oxidized during the STO growth. No oxide contribution is observed on the As3d core level because of the volatility of As oxides[14].

XPS core levels are clearly different for the sample grown without Ti pre-treatment. First, a small contribution (Fig.2(d)) appears in the Sr3d core level with a chemical shift of ~+1 eV with respect to the bulk position. The origin of this contribution has not been clearly established yet but seems to be related to the degree of crystallinity of the STO layer.

In Fig.2(e), no change is observed for the As3d core level as compared to the sample with Ti-treatment. In contrast, Ti3p(STO) component intensity is decreased by half and an additional component is observed at - 5.1eV from the Ti3p(STO) peak, close to the position expected for metallic Ti. Simultaneously, in Fig.2(f), a contribution appears in the Ga3d core level at -0.63eV from the Ga3d (GaAs) component. This simultaneity suggests that an alloy Ti_xGa_y is formed when no Ti-treatment is used prior to STO growth. This alloy is polycrystalline as shown by the RHEED pattern of Fig.1(b). The new proportion of the different components in Ga3d core level is 28.7% of GaAs, 21.3% of oxides and 50% of Ti_xGa_y. This indicates a competition between oxidation and formation of a polycrystalline metallic compound. Oxides are favored when a Ti pre-treatment is done, otherwise polycrystalline metallic compound.

These results highlight the advantage of using a Ti-treatment prior to STO growth on GaAs: this pre-treatment stabilizes the STO/GaAs interface and prevents the formation of a Ti_xGa_y polycrystalline compound which is damaging for epitaxial growth.

Integration of ferroelectric PZT on STO/GaAs

The complete structure (PZT/LSMO/STO/GaAs, including Pd top contacts for electrical characterization, see hereafter) was analysed by XRD (Fig.3). A general out-of-plane scan is displayed in Fig.3(a). Only 00l reflexions of the LSMO and PZT layers are detected, confirming that the oxide stack is single crystalline and purely (001)-oriented.Fig.3(b) shows a rocking curve recorded around the PZT 002 reflexion. The full width at half maximum (FWHM) of the peak is 2.2°. This value is comparable to that reported by Huang et al. for PZT grown by laser-MBE on STO template GaAs (1.5 to 2°)[11].

Palladium interdigital electrodes (including a contact pad and digits, see the inset of Fig.4) were deposited on the oxide heterostructure using a conventional lift-off process. Combined with the presence of the LSMO bottom electrode, the interdigitated surface electrode configuration allows for measuring the dielectric response of the stack avoiding a serial contribution of the GaAs substrate. Indeed, the dielectric constant of GaAs is much lower than the one of PZT so that the contribution of the semiconductor to the total capacitance would have masked that of PZT in a vertical measurement configuration. In addition, interdigitated electrodes also allow to measure possible dielectric in-plane contribution.

Three types of electrodes with equivalent total surface and consisting of respectively 16, 27 and 41 digits were fabricated. The surface of the digits is 3.2×10^{-8}, 2.7×10^{-8} and 2.05×10^{-8} m^2 for the 16, 27 and 41 digit electrodes, respectively. Fig.4 displays the C-V characteristic recorded on a 41-digit electrode at 1 MHz with a 30mV AC excitation using a HP4280A capacitancemeter. It is clearly butterfly-shaped, with a memory window of 1.1 V, demonstrating that the PZT layer is ferroelectric. The maximum capacitance C_{max} is 0.77 nF. The values of C_{max} recorded for the different electrode types are very close to each other within experimental uncertainties and do not scale the total digit surface. This suggests that the C-V curve in Fig.4 corresponds to the out-of-plane dielectric response of the PZT layer (electrical field lines cross the PZT layer along the out-of-plane direction and propagate laterally from one electrode to the neighbouring one through the low resistivity LSMO).

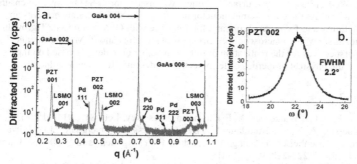

Fig. 3: (a) General XRD scan recorded on the complete heterostructure. The PZT layer is single crystalline and purely (001)-oriented. (b) Rocking curve recorded around the PZT 002 reflexion. The FWHM of the peak is 2.2°.

Indeed, the in-plane dielectric response is expected to scale the total digit surface as it consists in the parallel association of the individual capacitances formed by neighbouring digits. Oppositely, the out-of-plane dielectric response corresponds to a serial configuration where the capacitance scales the total electrode surface that does not depend on the electrode type in the present case. As a consequence, the C-V curve depicted in Fig.4 corresponds to the out-of-plane dielectric response of the PZT layer. The relative permittivity of the PZT layer extracted from these measurements is $\varepsilon_r = 164$. This value is amongst the highest reported for PZT layers of equivalent thickness[15,16] grown on Si.

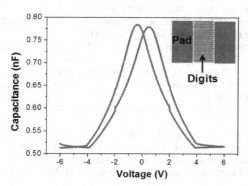

Fig. 4: C-V curves and its interdigited electrodes pattern.

CONCLUSIONS

Ferroelectric PZT was grown by PLD on STO/GaAs templates fabricated by MBE. A RHEED-controlled Ti-based GaAs surface treatment was used prior to STO growth to ensure crystal coherence across the oxide/semiconductor interface. The STO template presents an excellent cristallinity and forms an abrupt and flat interface with GaAs prior to PZT growth. PZT is ferroelectric, as shown by macroscopic electrical measurements and PFM experiments. These results open perspectives for the realization of micro-optoelectronic devices based on the combination of the physical properties of ferroelectric oxides and III-V semiconductors.

REFERENCES

1. R.A. McKee, F.J. Walker and M.F. Chisholm, Phys. Rev. Lett. **81**, 3014, (1998)
2. R. Droopad, Z. Yu, J. Ramdani, L. Hilt, J. Curless, C. Overgaard, J.L. Edwards, J. Finder, K. Eisenbeiser, J. Wang, V. Kaushik, B.-Y. Ngyuen, and B. Ooms, Journal of Crystal Growth **227-228**, 936, (2001)
3. Y. Liang, J. Kulik, T.C. Eschrich, R. Droopad, Z. Yu, and P. Maniar, Applied Physics Letters **85**, 1217, (2004)
4. A. Lin, X. Hong, V. Wood, A.A. Verevkin, C.H. Ahn, R.A. McKee, F.J. Walker, and E.D. Specht, Applied Physics Letters **78**, 2034, (2001)
5. S. Baek, J. Park, D. Kim, V. Aksyuk, R. Das, S. Bu, D. Felker, J. Lettieri, V. Vaithyanathan, S. Bharadwaja, N. Bassiri-Gharb, Y. Chen, H. Sun, C. Folkman, H. Jang, D. Kreft, S. Streiffer, R. Ramesh, X. Pan, S. Trolier-McKinstry, D. Schlom, M. Rzchowski, R. Blick, and C. Eom, Science **334**, 958 (2011)
6. S. Yin, G. Niu, B. Vilquin, B. Gautier, G.L. Rhun, E. Defay, and Y. Robach, Thin Solid Films **520**, 4595, (2012)
7. A.S. Borowiak, G. Niu, V. Pillard, G. Agnus, P. Lecoeur, D. Albertini, N. Baboux, B. Gautier, and B. Vilquin, Thin Solid Films **520**, 4604, (2012)
8. M. Scigaj, N. Dix, I. Fina, R. Bachelet, B. Warot-Fonrose, J. Fontcuberta, and F. Sánchez, Applied Physics Letters **102**, 112905, (2013)
9. G. Niu, S. Yin, G. Saint-Girons, B. Gautier, P. Lecoeur, V. Pillard, G. Hollinger, and B. Vilquin, Microelectronic Engineering **88**, 1232, (2011)
10. R. Contreras-Guerrero, J.P. Veazey, J. Levy, and R. Droopad, Applied Physics Letters **102**, 012907, (2013)
11. W. Huang, Z.P. Wu and J.H. hao, Appl. Phys. Lett. **94**, 032905 (2009)
12. L. Louahadj, R. Bachelet, P. Regreny, L. Largeau, C. Dubourdieu and G. Saint-Girons, submitted to Thin Solid Films
13. L. Ley, R.A. Pollack, F.R. McFeely, S.P. Kowalczyk, D.A. Shirley, Phys. Rev. **B 9**, 600 (1974)
14. C.W. Wilmsen, Thin Solid Films, **39**, 105 (1976)
15. D. Isarakorn, D. Briand, S. Gariglio, A. Sambri, N. Stucki, J. Triscone, F. Guy, J. Reiner, C. Ahn, and N. de Rooij, Smart Materials Research 2012, 426048, (2012)
16. H. Fujisawa, S. Nakashima, K. Kaibara, M. Shimizu and H. Niu, Jap. J. Appl. Phys. **38**, 5392, (1999)

Mater. Res. Soc. Symp. Proc. Vol. 1675 © 2014 Materials Research Society
DOI: 10.1557/opl.2014.876

Multilayer BiFeO₃/PbTiO₃ Multiferroic Ceramic Composites Prepared by Tape Casting

Guoxi Jin, Jianguo Chen and Jinrong Cheng[*]
School of Materials Science and Engineering, Shanghai University, 200444, P.R. China
[*]corresponding author: jrcheng@shu.edu.cn

ABSTRACT

The $BiFeO_3$ (BFO) / $PbTiO_3$ (PT) multiferroic ceramic composites with multilayered structure were prepared from orderly laminated BFO and PT tapes by tape casting method. The dielectric constant ε_r, loss $tan\delta$, remnant polarization P_r and field-induced strain of BFO/PT ceramic composites were 140 (1 kHz), 5% (1 kHz), 12 $\mu C/cm^2$ (at 80 kV/cm) and 0.06% (at 80 kV/cm) respectively, which were comparable to those pure BFO ceramics and BFO-based solid solutions

INTRODUCTION

In recent years, multiferroic materials, which exhibit simultaneous ferroelectric and ferromagnetic behaviors, have been receiving considerable attention due to their interesting magnetoelectric effects promised for potential multifunctional application in memories, transducers, sensors, and switching devices, etc. [1–4].

Among the few room-temperature single-phase multiferroic materials reported, $BiFeO_3$ (BFO) is a promising candidate due to its high Curie (T_c 1103 K) and Neel (T_N 643 K) temperatures [5-6]. However, its low resistivity and weak ferromagnetic property at room temperature makes it difficult to be used in practical applications [7]. Consequently, a variety of improvements including formation of solid solution (with $PbTiO_3$, $BaTiO_3$ or any other ABO_3, etc.) and rare-earth elements doping has been widely and deeply researched to ameliorate those defects above [8-10]. Our previous work showed that introducing $PbTiO_3$ into $BiFeO_3$ to form BFPT solid solution enhanced insulating and multiferroic properties. And with the La, Mn, Ga doping, the coercive field, dielectric constant and loss of BFPT solid solution can be sequentially decreased [11-14].

However, there is less work on the BFO-based composites such as multilayered structure, which is expected to improve the ferroelectric and piezoelectric properties, has been employed to study the ferroelectric, piezoelectric and magnetic properties. Tape casting method is widely used to prepare multilayer structure materials with exact scale control. It owns the great advantage in forming large-area, thin and flat ceramic parts and excellent convenience in composites structure design. With the great progress in tape casting preparation on multiferroic ceramics [15], now the BFO-based composites with multilayered structure can be easily achieved.

In the present work, $PbTiO_3$ (PT) and $BiFeO_3$ (BFO) tapes were obtained separately by optimized tape casting method based on our previous work. And then, these two kinds of ceramic tapes were laminated orderly to form multilayered composites instead of traditional solid solutions. Dielectric, ferroelectric and piezoelectric properties of BFO/PT composites with multilayered structure were studied.

EXPERIMENT

Analytical-grade powders of Bi_2O_3, TiO_2, Fe_2O_3 and PbO were used as starting materials to synthetize $BiFeO_3$ (BFO) and $PbTiO_3$ (PT) respectively. Firstly, they were blended and adequately ball-milled with deionized water for 24 h. Then, the mixture was dried and calcined at 750 °C for 4 h. The processes of ball-milling and calcination were repeated twice. The prepared powders were mixed in ethanol with TEA (dispersant) for 2 h first. Then PVB (binder) and PEG-DBP (plasticizer) were added into the slurry, and followed by mixing for another 12 h to obtain the fine and uniform slurry. The slurry was then casted on a plate glass surface and dried at room temperature in the open air without blowing. The green sheets of BFO and PT were punched, laminated in order and cold isostatic pressed at 200 MPa for 2 min to obtain composite disks of 12 mm in diameter. The pressed tablets were heated in air at 600 °C for 3 h to remove all organic additives and then sintered at 880 °C in air for 3 h. Finally, the composite ceramics were polished and coated with sliver electrodes on both sides. The crystallographic phases of prepared BFO, PT powders and ceramic composites were characterized by X-ray diffraction (Rigaku D/Max 2200, Japan). The specimen cross morphology was observed by the scanning electron microscopy (Hitachi S3400, Japan). The dielectric, ferroelectric properties and field-induced strain were measured by a precision impedance analyzer (Agilent HP4294A, USA), a fotonic sensor (MTI 2000, USA) and a ferroelectric measurement system (Radiant Premier II, USA), respectively.

RESULTS AND DISCUSSION

Figure 1 shows the $BiFeO_3$, $PbTiO_3$ tapes and single slices of the tapes. The thickness of the tapes was controlled at 500 μm in casting process while shrinking to about 100 μm after drying in air. The diameter of the slices is designed at 12 mm. They were laminated in order with 20 layers to form a pressed bulk. The green bodies underwent the organic additives removal, cold isostatic pressing procedures and sintering process. As a result, the diameter shrunk from 12 mm to about 10 mm.

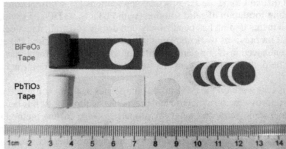

Figure 1. Photographs of the $BiFeO_3$, $PbTiO_3$ tapes and as-cut slices

The X-ray diffraction patterns of prepared $BiFeO_3$, $PbTiO_3$ powders and BFO/PT ceramic composites are shown in Figure 2. The patterns show that pure $PbTiO_3$ powders with no impurity phases while $BiFeO_3$ powders with few impurity peaks (indicated by star sign) were obtained respectively. Though highly pure $BiFeO_3$ powders were not synthetized, the second phases matter disappeared gradually in composite with the temperature increasing. The Bragg reflections of the BFO/PT composites phase peaks do not shift but some new phases appeared, indicating that there must be a little interface reaction between BFO and PT occurred during the higher temperature sintering.

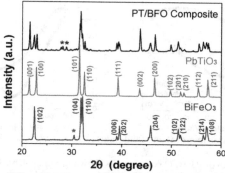

Figure 2. X-ray diffraction patterns of $BiFeO_3$, $PbTiO_3$ and PT/BFO composites

SEM investigation performed on the PT/BFO composite specimen fracture surface are presented in Figure 3. The green body of PT and BFO tapes lamination in order is shown in Figure 3(a) which indicated the multilayered structure clearly. Moreover, the ordered arrangement was not destroyed by the interface reaction during the sintering. In Figure 3(b), it shows the interface between PT and BFO phases in sintered ceramic composites. It is obvious that there are two types of grains coexistence which is identical with the X-ray diffraction of PT/BFO composites. In addition, the compact combination between PT and BFO layers indicates the well sintered at 880 °C for 3 h which is much lower than the BFPT solid solution [15].

Figure 3. SEM on fracture surface of PT/BFO composites

The polarization-electric field (P–E) hysteresis loops at room temperature for PT/BFO composite ceramic is given in Figure 4(a), the composite exhibits an unsaturated hysteresis loop which is rounded at the highest field. This may be due to increase of leakage current in the system. It was also found that the remnant polarization and coercive field were 12 $\mu C/cm^2$ and 50 kV/cm, respectively. The observed polarization and coercive field were larger than the pure BFO ceramcis reported values [16], which probable is due to the enhanced insulated property, excellent densification in composite and the coexistence of PT phase.

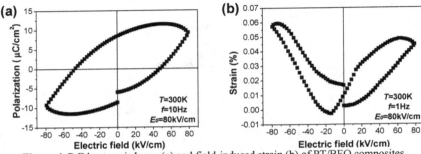

Figure 4. P-E hysteresis loops (a) and field-induced strain (b) of PT/BFO composites

The field-induced strain of PT/BFO composite is shown in Figure 4(b). The bipolar strain of the composite reaches 0.06% at 80 kV/cm (1 Hz) which is comparable to the well-prepared BFO ceramics (strain of 0.08% at 130 kV/cm) reported in another paper [17]. This is the first time to report the field-induced strain of a multilayered composite ceramic based on BFO materials

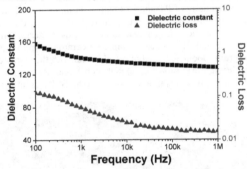

Figure 5. Dielectric constant and loss of PT/BFO composites as a function of frequency

Figure 5 shows the dielectric constant and loss of PT/BFO composites prepared by casting tapes as a function of frequency from 100 Hz to 1 MHz. The dielectric constant of the specimen is 160 at 100 Hz with the loss of 12.4% while decreasing to 140 at 1 kHz with the loss of 5%.

Temperature dependent dielectric constant and loss of PT/BFO composites measured at different frequencies (10^2, 10^3, 10^4, 10^5 and 10^6 Hz) are displayed in Figure 6. It is found obvious that a phase-transition peak is only observed at about 600 °C. That may be caused by the thermal mismatch stress, which occurred between the interfaces, averages the transition temperatures between BiFeO$_3$ and PbTiO$_3$. Furthermore, the PT/BFO composites reveal a stable change with the increasing temperature until 300 °C, reflecting the improved high temperature resistivity.

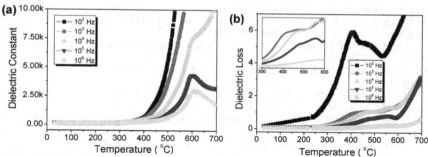

Figure 6. Dielectric constant (a) and loss (b) of PT/BFO composites as a function of temperature

CONCLUSIONS

The PbTiO$_3$ (PT) / BiFeO$_3$ (BFO) ceramic composites with multilayered structure were prepared by optimized tape casting method. Properties of PT/BFO composites including ferroelectricity, field-induced strain, dielectric constant and loss as a function of frequency and temperature were discussed in detail. The laminated ceramic composites exhibit comparable electrical performances to those traditional pure BFO ceramics and BFO-based solid solutions. In addition, with the tape casting method, a variety of interesting and special structure can be easily obtained. It is an access to research the effects of composition and structure design to the electrical properties of advanced materials.

ACKNOWLEDGMENTS

This work was financed by the National Nature Science Foundation of China under grant No.51302163.

REFERENCES

1. J. Ma, J. M. Hu, Z. Li, C.W. Nan, *Adv. Mater.*, **23**, 1062–1087 (2011).
2. C. C. Zhou, B. C. Luo, K. X. Jin, X. S. Cao, C. L. Chen, *Solid State Commun.*, **150**, 1334–1337 (2010).
3. Z. H. Dai, Y. Akishige, *Ceram. Int.*, **38**, S403–S406 (2012).
4. M. Rawat, K. L. Yadav. *J. Alloy. Compd.*, **597**, 188–199 (2014)
5. J. Wang, J. B. Neaton, H. Zheng, V. Nagarajan, S. B. Ogale, B. Liu, D. Viehland, V. Vaithyanathan, D. G. Schlom, U. V. Waghmare, N. A. Spaldin, K. M. Rabe, M. Wuttig, R. Ramesh, *Science*, **299**, 1719 (2003).
6. D. H. Wang, L. Yan, C. K. Ong, Y. W. Du, *Appl. Phys. Lett.*, **89**, 182905 (2006).

7. D. I. Woodward, I. M. Reaney, R. E. Eitel, C. A. Randal, *J. Appl. Phys.*, **94**, 3313 (2003).
8. K. Singh, N. S. Negi, R. K. Kotnala, M. Singh, *Solid State Commun.*, **148**, 18–21 (2008).
9. R. A. Gotardo, I. A. Santos, L. F. Cotica, E. R. Botero, D. Garcia, J. A. Eiras, *Scripta Mater,.* **61**, 508–511 (2009).
10. T. Leist, T. Granzow, W. Jo1, J. Rodel, *J. Appl. Phys.*, **108**, 014103 (2010).
11. J. R. Cheng, R. Eitel, L. E. Cross, *J. Am. Ceram. Soc.*, **86** [12], 2111–2115 (2003).
12. J. R. Cheng, N. Li, L. E. Cross, *J. Appl. Phys.*, **94**, 5153 (2002).
13. G. Y. Shi, J. G. Chen, L. Zhao, S. W. Yu, J. R. Cheng, L. Hong, G. R. Li, *Curr. Appl. Phys.*, **11**[3], S251-S254 (2011).
14. G. Y. Shi, L. Feng, S. D. Bu, W. Ruan, G. Li, T. Yang, J. R. Cheng, *Smart Mater. Struct.*, **21**, 065009 (2012).
15. G. X. Jin, J. G. Chen, S. D. Bu, D. L. Wang, R. Dai, J. R. Cheng, *Mater. Res. Soc. Symp. Proc.*, **1547**, 61-66 (2013).
16. Y. P. Wang, L. Zhou, M. F. Zhang, X. Y. Chen, J. M. Liu, Z. G. Liu, *Appl. Phys. Lett.*, **84**, 1731 (2004).
17. T. Rojac, M. Kosec, D. Damjanovic, *J. Am. Ceram. Soc.*, **94** [12], 4108–4111 (2011).

Mater. Res. Soc. Symp. Proc. Vol. 1675 © 2014 Materials Research Society
DOI: 10.1557/opl.2014.778

Effect of Ca and Ag doping on the functional properties of BiFeO$_3$ nanocrystalline powders and films

Gina Montes Albino[1], Oscar Perales-Pérez[3], Boris Renteria-Beleño[3] and Yarilyn Cedeño-Mattei[3]

[1]Department of Mechanical Engineering, University of Puerto Rico at Mayagüez P.O. Box 9045, Mayagüez, PR, 00681-9045 USA.
[2]Department of Physics, University of Puerto Rico at Mayagüez, Mayagüez, PR, 00980, USA.
[3]Department of Engineering Science and Materials, University of Puerto Rico at Mayagüez, Mayagüez, PR, 00680-9044, USA.

ABSTRACT

The present work addresses the systematic evaluation of the influence of the incorporation of dopant species (Ca^{+2}, Ag^{+1}) on the structural and functional properties of bismuth ferrite (BFO) nanocrystalline powders and films. Pure and doped BFO powders and thin films were synthesized by a modified sol-gel method. The concentration of the doping species varied from 0 up to 7 at %. The development of the host BFO structure was confirmed by XRD analyses of samples annealed at 700°C for one hour in air and nitrogen atmosphere. Thicknesses of films varied between 80 and 200 nm, depending on the concentration of Ca^{+2} species. Doped BFO exhibited a magnetic behavior that turned from paramagnetic into ferrimagnetic with the increase of Ca^{+2} concentrations.

INTRODUCTION

Proper-primary type BiFeO$_3$ (BFO)-based multiferroics are rhombohedrally distorted acentric structure (space group R3c) that exhibit G-type antiferromagnetic order with a long-periodicity spiral below the Neel temperature of 643 K and ferroelectricity below 1103 K. The ferroelectricity of BiFeO$_3$ is due to Bi^{3+} 6s^2 lone-pair distortions, whereas the residual moment of the canted Fe^{3+} spin structure results in weak ferromagnetism. Recently, perovskite-type transition metal oxides ABO$_3$ are of great interest because of their magnetic, dielectric, and transport properties that emerge from the coupling of spin, charge, and orbital degrees of freedom. A-site substitutions of trivalent (La^{3+}, Nd^{3+}, or Sm^{3+}) [1-6] or divalent (Ba^{2+}, Pb^{2+}, Sr^{2+}, or Ca^{2+}) [7-10] species for Bi^{3+} and B-sublattice dopings with V^{5+}, Nb^{5+}, Mn^{4+}, Ti^{4+}, or Cr^{3+} ions, have recently been investigated in order to improve the magnetoelectric coupling [11-15]. BFO perovskite type, proper and lead-free multiferroic material can find potential and promising applications in the development of multifunctional devices, solar energy devices, ferroelectric random access memory and spintronics. These applications rely on the observed (anti)ferroelectricity, (anti)ferromagnetism, and ferroelasticity behaviors in a single crystalline phase. On this basis, the present work will attempt to find a suitable combination of deposition parameters, via a modified sol-gel approach, for improving multifunctional BFO properties in powders and films. If oxygen vacancies are generated during the material forming stage, structural, magnetic and electric transitions could be tuned and hence, optimized. Therefore, the annealing stage of the precursor solids will be carried out in a close-system in air or nitrogen atmospheres. Furthermore, the expected variation in the structural and functional properties of BFO in presence of different dopant species (A-site doping with Ca^{+2} and Ag^{+1} cations) will also be evaluated; doping with a divalent cation would induce a p-type semiconductor behavior favoring magnetic transition.

EXPERIMENTAL

Materials

Pure and doped bismuth ferrite thin films were synthesized in acetic acid medium. Bismuth Nitrate [$Bi(NO_3)_3.5H_2O$, purity 99.9%], Iron Nitrate [$Fe(NO_3)_3.9H_2O$, purity 99.9 %], Calcium Nitrate [$Ca(NO_3)_2.4H_2O$ purity 99%] and Silver Nitrate [$AgNO_3$, purity 99.99 %], salts. Suitable weights of $Bi_{1-x}M_xFeO_3$ (M= Ca or Ag) salts are used to achieve the desired atomic percentages, 'x', according to the stoichiometry.

Synthesis of powders and thin films

Powders synthesis started with the dissolution of precursor salts in acetic acid for 1 hour followed by the dropwise addition of the resulting solution on petri dish, and finally dried in air for 12 hours at 80°C. Thin films were synthesized by first dissolving the precursor salts in acetic acid and contacted with glycol (1mL of glycol added to 4mL of the precursor solution) in order to control the viscosity. The homogeneous solution was added dropwise onto a clean Si (100) substrate and spin-coated at 3000 rpm for 20 s. After each coating cycle, produced films are pre-dried for 5 minutes at two different temperatures (150°C and 250°C) in order to remove organic residuals. These spin-coating/drying cycles were repeated for twenty times to thick the films. Powders and spin-coated films were then thermally treated in air or nitrogen for one hour at 700°C. The heating rate was 5 °C/min in all experiments.

Materials characterization

The structure of the synthesized powders and films was determined using a Siemens D5000 X-Ray Diffractometer (XRD) with Cu-Kα radiation. The room-temperature magnetic hysteresis (M-H) loops were measured using a Lake Shore 7410 Vibrating Sample Magnetometer (VSM).

DISCUSSION

X-ray diffraction analyses

Pure and Ca doped BFO powders and films

XRD spectra for Ca-BFO samples produced after the thermal treatment stage are shown in figures 1 through 3. Figures 1 and 2 show the Ca-BFO(0 - 7 at % Ca) powders annealed in air and nitrogen atmosphere, respectively. As observed in figures 1a and 2a, main XRD peaks corresponding to BFO phase coexisted with low intensity peaks in the 25° - 30° 2θ range, which are attributed to small amount of impurity phases ($Bi_2Fe_4O_9$ or $Bi_{25}FeO_{39}$ intermediate oxides). The same XRD patterns suggest that a suitable Ca doping could be conducive to the inhibition of the formation of the impurity phases (7 at.% and 2 at.% of Ca in air and nitrogen atmosphere, respectively). Average crystallite sizes for varied from 42 to 26nm and 42 to 33nm, in powders annealed in air or nitrogen, respectively. Figures 1b and 2b show the magnification of the peak corresponding to the (204) plane; a slight but evident shift towards lower 2θ values was observed and suggested the actual dopant incorporation in the BFO host structure. The ionic radii of Ca^{+2} is larger (1.26Å) than Bi^{+3} (1.17Å); this difference would explain the corresponding expansion in the unit cell as suggested by the shift of the XRD peaks[16-17]. Figure 3 shows the XRD pattern corresponding to the Ca-doped BiFeO$_3$ thin film deposited onto Si (100) by spin coating process. The film thickness was 200 nm. Only the main characteristic peaks corresponding to the BFO structure were observed. Thee inhibition on the formation of the impurities could be attributed to substrate effects as well as the suitable calcium content.

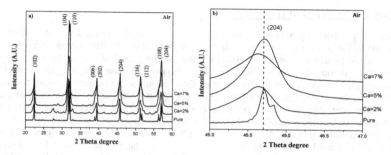

Fig 1. XRD patterns of: (a) pure and Ca- BFO powders annealed in air. (b) magnification of the angular region around the (204) peak; a slight shift to lower 2θ values as Ca- dopant content increased from 0 to 7 at.% was observed.

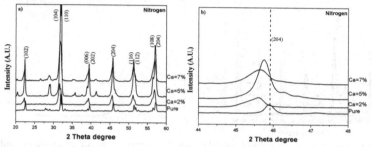

Fig 2. XRD patterns of: (a) pure and Ca-BFO powders annealed in nitrogen atmosphere and, (b) magnification of the angular region around the (204) peak.

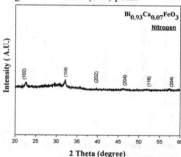

Fig 3. XRD pattern corresponding to 20-layers Ca-BFO/Si (100) thin film. The Ca content was set at 7 at %.

Pure and Ag-doped BFO powders

XRD spectra for Ag-BFO powders formed under air and nitrogen atmospheres are shown in figures 4 and 5. The main peaks corresponding to the BFO host structure that co-existed with impurity Bi-oxide phases were detected. The electronic decompensation due to the substitution of trivalent Bi^{+3} by monovalent Ag^+ would have altered oxygen diffusion promoting the formation of other oxide phases. Despite of the presence of these impurities, a remarkable shift towards lower diffraction angles was observed for the BFO peaks (Figure 4b and 5b) in presence of 2 at. % of Sag; it was interesting to observe that the (204) peak was slightly shifted towards higher 2θ values when the Ag content was increased up to 7 at.%. The expansion of the unit BFO unit cell due to the incorporation of large Ag ions (1.42 Å) could be related to this displacement of the diffraction peaks of the host BFO structure. In turn, the average crystallite sizes for Ag-BFO powders treated in air and nitrogen decreased 40 to 26 nm (annealed in air) and 40 to 37 nm (annealed in nitrogen), when the dopant content increases from 0 up to 7%.

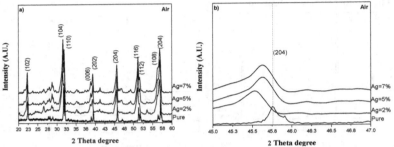

Fig 4. XRD patterns of: (a) pure and Ag-BFO powders annealed in air atmosphere, and (b) magnification of the angular region around the (204) peak.

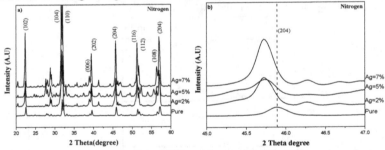

Fig 5. XRD patterns of: (a) pure and Ag-BFO powders annealed in nitrogen atmosphere, and (b) magnification of the angular region around the (204) peak.

Room-temperature M-H measurements
Pure and Ca doped BFO powders and films

Room-temperature M-H loops of pure and Ca-BFO powders annealed for one hour at 700°C in air and nitrogen are shown in figures 6 a and b, respectively. As evident, doping of BFO with Ca drastically changed the magnetic behavior from paramagnetic to ferrimagnetic[18]. A rising trend in coercivity was also observed when the Ca concentration was increased up to 7%. The corresponding coercivity values varied from 10 to 16 Oe when annealed in air (figure 6a) and from 5 to 35 Oe when the annealing was performed in nitrogen (figure 6b) atmosphere. In turn, figure 7 shows the M-H loops for thin films of Ca (7 at. %)-BFO/ Si(100) annealed at 700°C in nitrogen atmosphere. Maximum magnetization values was dependent on the thickness vary from 80 to 200nm depending of different coatings in the film. Coercivity values are 40, 33, and 23 Oe for 20, 15, and 10 layers, respectively.

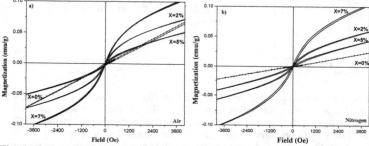

Fig 6. Room-temperature M-H loops corresponding to Ca-doped bismuth ferrite powders synthesized at different calcium atomic percentages, 'x', annealed at 700°C for 1hour in: (a) air and (b) nitrogen.

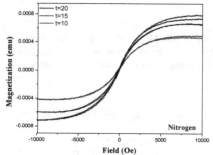

Fig 7. Room-temperature M-H loops for Ca (7 at.%)-BFO/Si(100) films with different thicknesses ('t' is the numbers of layers used in the precursor film preparartion. The magnetic field was applied perpendicular to the film plane. The coercivity was as high as 40 Oe for 20 layer thick film (thickness = 200nm) doped at 7 at % Ca.

Pure and Ag doped BFO powders

Figure 8 shows the M-H loops of Ag-BFO powders annealed in air. The hysteresis loops exhibit the same tendency as in Ca-BFO powders, i.e., the material changed from paramagnetic into ferri/ferromagnetic after doping with Ag species. Coercivity was as high as 97 Oe for the BFO doped with 7 at. % of Ag. The observed effect of the Ca and Ag dopants on the magnetic nature of the BFO host could be attributed to apparent incorporation of Ca and Ag cations into the BFO lattice as well as to the probable p-type semiconductor behavior[19, 20].

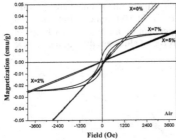

Fig 8. Room-temperature M-H loops for Ag- doped bismuth ferrite powders synthesized at different silver atomic percentages, 'x', annealed at 700°C by 1h in air atmosphere.

CONCLUSIONS

Pure and Ca- and Ag-BFO nanocrystalline powders and thin films were synthesized using a modified sol-gel technique. X-ray diffraction analyses confirmed the formation of well crystallized cubic rhombohedral BFO phase. BFO single phase formations were obtained for 7 and 2 at % of Ca- doped BFO powders in air and nitrogen atmosphere, respectively. M-H measurements of the Ca-BFO powders and films and the Ag-BFO powders evidenced the strong influence of the type and amount of dopants on the magnetic characteristics of the host BFO structure; depending on the dopant amount, the material drastically changed from a paramagnetic into a ferromagnetic ones.

ACKNOWLEDGMENTS

This material is based upon work supported by the DOE-Grant No FG02-08ER46526. Special thanks to Brian Ferrer and Walter Guzman who helped with the synthesis process. Particular acknowledgemt to the support from Dr. Ram Katiyar's group from the Speclab Laboratory at UPR-Rio Piedras Campus.

REFERENCES

[1] Zhang, S.-T.; Pang, L.-H.; Zhang, Y.; Lu, M.-H.; Chen, Y.-F. J. Appl. Phys. 2006, 100, 114108.

[2] Yuan, G. L.; Or, S. W.; Chan, H. L. W. J. Phys. D: Appl. Phys. 2007,40, 1196–1200.

[3] Mathe, V. L.; Patankar, K. K.; Patil, R. N.; Lokhande, C. D. J. Magn. Magn. Mater. 2004, 270, 380–388

[4] Yuan, G. L.; Or, S. W.; Liu, J. M.; Liu, Z. G. Appl. Phys. Lett. 2006,89, 052905.

[5] Mathe, V. L. J. Magn. Magn. Mater. 2003, 263, 344–352.

[6] Yuan, G. L.; Or, S. W. J. Appl. Phys. 2006, 100, 024109.

[7] Wang, D. H.; Goh, W. C.; Ning, M.; Ong, C. K. Appl. Phys. Lett.2006, 88, 212907.

[8] Li, J.; Duan, Y.; He, H.; Song, D. J. Alloys Compd. 2001, 315, 259–264.

[9] Khomchenko, V. A.; Kiselev, D. A.; Vieira, J. M.; Kholkin, A. L.;Sa, M. A.; Pogorelov, Y. G. Appl. Phys. Lett. 2007, 90, 242901.
[10] Kothari, D.; Reddy, V. R.; Gupta, A.; Sathe, V.; Banerjee, A.; Gupta,S. M.; Awasthi, A. M. Appl. Phys. Lett. 2007, 91, 202505.
[11] Yu, B.; Li, M.; Liu, J.; Guo, D.; Pei, L.; Zhao, X. J. Phys. D: Appl. Phys. 2008, 41, 065003.
[12]Jun, Y.-K.; Moon, W.-T.; Chang, C.-M.; Kim, H.-S.; Ryu, H. S.; Kim, J. W.; Kim, K. H.; Hong, S.-H. Solid State Commun. 2005, 135, 133–137.
[13] Yang, C. H.; Koo, T. Y.; Jeong, Y. H. Solid State Commun. 2005,134, 299–301.
[14] Santos, I. A.; Grande, H. L. C.; Freitas, V. F.; de Medeiros, S. N.;Paesano, J. A.; Co´tica, L. F.; Radovanovic, E. J. Non-Cryst. Solids 2006, 352, 3721–3724.
[15] Qi, X.; Dho, J.; Tomov, R.; Blamire, M. G.; MacManus-Driscoll, J. L. Appl. Phys. Lett. 2005, 86, 062903.
[16] B. Bhushana,, D. Das, A. Priyamc N.Y. Vasanthacharya, S. Kumar; Materials Chemistry and Physics 135 (2012) 144-149
[17] Ji-Zhou Huang, Yang Shen, Ming Li,and Ce-Wen Nan: JOURNAL OF APPLIED PHYSICS 110, 094106 (2011).
[18] D. Rubia, F.G. Marlascaa, M. Reinosoa,b, P. Bonvilled, P. Levy; Materials Science and Engineering B 177 (2012) 471–475
[19] M.A. Ahmeda, S.F. Mansour , S.I. El-Deka, M. Abu-Abdeenc; Materials Research Bulletin 49 (2014) 352–359
[20] Nahum Maso and Anthony R. West; Chem. Mater. 2012, 24, 2127-2132

Mater. Res. Soc. Symp. Proc. Vol. 1675 © 2014 Materials Research Society
DOI: 10.1557/opl.2014.802

Magnetic and optical properties of Mn-doped SnO$_2$ films

S.Sujatha Lekshmy, Anitha V.S, K. Joy*
*Thin film Lab, Post Graduate and Research Department of Physics,
Mar Ivanios College, Thiruvananthapuram 695015.*
*Corresponding author e-mail: jolly2jolly@gmail.com

ABSTRACT

Magnetic nanoparticles have drawn much attention due to their potential in magnetic recording as well as many biological and medical applications such as magnetic separation, hyperthermia treatment, magnetic resonance contrast enhancement and drug delivery. The magnetic fields generated by these nanoparticles can be used for diagnostics in Magnetic Resonance Imaging (MRI) etc. Manganese doped tin dioxide (SnO$_2$:Mn) possess interesting physical and chemical properties. The physical and chemical properties of the particles themselves like the size, shape, crystallinity and composition, will control the magnetic properties and response of the particles to magnetic fields. Our work is rooted to control the properties of the particles as well as tailor their magnetic properties for specific applications. In this study, SnO$_2$: Mn films with different Mn doping concentrations (0-3 mol%) were deposited on the glass substrates by sol-gel dip coating technique. XRD patterns shows tetragonal structure for all the SnO$_2$:Mn films and crystallite size decreased as Mn doping concentration increased from 0 - 3 mol%. The magnetic property shows that pure SnO$_2$ film is diamagnetic and 1- 3 mol% SnO$_2$:Mn films posses room temperature ferromagnetism. The optical properties of the films revealed that transmittance of the films decreased with increase in Mn doping concentration. The optical energy band gap values (3.55 eV-3.71 eV) increased with the increase in Mn doping concentrations. Such SnO$_2$:Mn films with structural, optical and magnetic properties can be used as dilute magnetic semiconductors.
Keywords: Tin oxide, Thin films, Sol gel processing, X-ray diffraction, Magnetic properties

1. INTRODUCTION

The rapidly developing field of spin electronics requires a semiconducting room temperature ferromagnet for incorporation in spin-electronic devices. Among the necessary requirements in order to have a spintronic device , is the need of an efficient electrical injection of spin-polarized carriers (spin injection) into the semiconductor. Tin dioxide (SnO$_2$) is an attractive semiconductor for the fabrication of diluted magnetic semiconductors (DMS) because of its excellent optical transparency, native oxygen vacancies and high carrier density . In the quest for materials which involve both the charge and spin of electrons in a single substance, studies have also paid considerable attention on SnO$_2$-based DMS realized through transition-metal (TM) doping. During the past few years, the DMS in which a small fraction of atoms/ions is magnetic, have attracted considerable attention from both the fundamental as well as the application points of view. This is primarily due to the possibility that, in these DMSs, the two degrees of freedom (spin and charge) can be independently tuned to realize multifunctional devices. In such devices, inducing stable room temperature ferromagnetism (RTFM) is one of the necessary prerequisites to realize any practical application of these materials[1]. Doping enhances the properties of semiconductors by providing a powerful method to control their optical, magnetic, transport, and spintronic properties [2,3]. The optoelectronic properties such as photoluminescence and optical band gap of SnO_2 can also be improved by impurity doping. Among the transition metals, much emphasis is being put on manganese (Mn) due to its large equilibrium solubility and nearly the same ionic radii compared to Sn^{4+} ion for substitution.

In recent days, thin films of SnO_2 : Mn have become an integral part of modern electronic technology. Tailoring the physical properties and adding new functionalities to the existing SnO_2 semiconductors by altering the structure, composition and particle or grain size are the new approaches in advancing the current applications of SnO_2 semiconductor materials.

The optical properties with size effect of nanocrystalline semiconductors have been a subject of great interest in recent years. When the particle size of a semiconductor reduces to a few nanometer in range, the ratio of surface atoms to those with the interior enhances the surface properties of the material. Due to these effects, band gap of the semiconductor nanoparticles gets modified, which makes a lot of changes in their electronic properties when compared to the bulk [4]. This stimulated a great interest in both basic and applied research in semiconductor oxide nanomaterials.

In this work, we present the effect on the structural, magnetic and optical properties by doping different concentrations of Mn in SnO_2 films produced by sol gel dip coating technique.

Sol–gel fabrication has gained much interest because of its simplicity, low processing temperature, stoichiometry control and its ability to produce uniform, chemically homogenous films over large areas that can provide integration with other circuit elements [5]. The main advantage of the sol-gel process is the ability to form inorganic structures at relatively low temperature.

2. EXPERIMENTAL DETAILS

SnO_2: Mn films were prepared by the sol–gel dip coating method. The following procedure was adopted for the preparation of the films. The sol was prepared by dissolving (0.5 mol) of $SnCl_2 \cdot 2H_2O$ in 50 ml absolute ethanol (99.7 %). The mixture was well stirred and refluxed at 80 °C for 2 h. Manganese chloride ($MnCl_2 \cdot 4H_2O$, 99 %) was then added into the solution as an Mn precursor with Mn concentration varying from 1 – 5 mol%. This solution was refluxed at 80 °C for 4 h, for the homogeneous mixing of the solution and then aged in air for 48 h. The clear homogenous solution was used for dip coating procedure. The SnO_2: Mn films were deposited on glass substrates, which were ultrasonically cleaned with acetone. The withdrawal speed of the substrate from the coating is 90 mm/min. Then the films were dried at 120 °C for 30 minutes, to evaporate the solvent and remove organic residuals. This coating and drying procedure was repeated several times to produce layers of solution on substrate. Then for crystallization, the samples were annealed in a furnace at 500 °C for 1 h in air. Upon annealing, the thin films were ready to be analyzed for its characteristics.

Crystallization phase of the SnO_2: Mn films were characterized by X-ray diffraction (XRD) using X-ray diffractometer (Model—PW 1710 PHILIPS). The magnetic properties were measured using the VSM (Lakeshore VSM 7410). The spectral transmittance of the films were recorded as a function of wavelength, in the range 300–900 nm, using JASCO V-550 UV–vis Spectrophotometer. The film thickness and optical band gap values were determined using Swanepoel's envelope method [6]. Photoluminescence (PL) spectra were recorded by using a Fluorescence spectrometer (Perkin-Elmer Model-LS55) with a 40 W xenon lamp as excitation source and 5 nm emission slit width.

3 RESULTS AND DISCUSSIONS

3.1 X-RAY DIFFRACTION STUDIES

XRD patterns of Mn doped SnO_2 (SnO_2: Mn) films deposited on glass substrate for different Mn doping concentration are shown in Figure 1. The XRD data of the undoped SnO_2 films revealed peaks at $26.6°$, $33.83°$, $38.08°$, $51.8°$, $54.8°$, $57.8°$, $64.9°$ and $66.03°$ corresponding to the (110), (101), (200), (211), (220), (002), (112) and (301) planes respectively. In the XRD pattern of SnO_2: Mn films, the peaks at $26.6°$, $33.83°$ and $51.8°$ corresponding to (110), (101) and (211) planes respectively were observed. All the prominent peaks in the pattern correspond to the tetragonal structure of SnO_2 and are indexed on the basis of JCPDS file no. 41-1445. No peaks corresponding to either Mn or Mn oxides appear in the diffraction pattern of the films, indicating that no detectable secondary phase exists in SnO_2: Mn films. This implies that the transition metal ions got substituted at the Sn site without changing the tetragonal structure of SnO_2. From the XRD patterns, it was observed that, on increasing the concentration of Mn, the intensity of the SnO_2 peaks were decreased. The decrease in peak intensities is basically due to the replacement of Sn^{4+} ions with Mn^{3+} ions in the lattice of SnO_2 film. This process leads to the movement of Sn^{4+} ions in the interstitial sites and also an increase in the disorder [7]. Hence, doping of Mn in SnO_2 not only reduces the crystallite size but also degrades the crystallinity of the material.

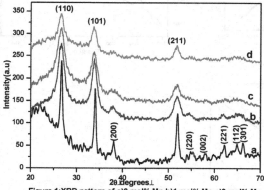

Figure 1:XRD pattern of a)0 mol% Mn b)1 mol% Mn c)3 mol% Mn d)5 mol% Mn in Mn doped SnO_2 films.

The average crystallite size of the SnO_2 thin films were calculated using Scherrer's equation [8].Increase in Mn content makes the full widths at half maximum (FWHM) of diffraction peaks to become broader, indicating reducing trend in the crystallite size. It can be observed from Table 1 that, the grain size of SnO_2 : Mn nanoparticles in the film reduced from 11.38 nm to 6.09 nm as a result of increase in Mn content from 0 mol% to 5 mol%. This indicates that the presence of Mn ions in SnO_2, prevented the growth of crystal grains. The presence of Mn ions in the crystallographic structure increases the formation of oxygen vacancies as required by the charge balance. This effect is in conjunction with the smaller ionic radius of Mn^{3+} ion (0.65 Å) in comparison to Sn^{4+} ion (0.69 Å). This can disturb long range crystallographic ordering and hence, reducing the crystallite size.

3.2 MAGNETIC STUDY

Room temperature magnetic measurements of the SnO_2: Mn films were carried out in the field range of ±15kOe. Fig 2(a-d) shows the M-H curves of pure and Mn doped SnO_2 films. The magnetization measurement for undoped SnO_2 sample exhibits diamagnetic nature (Fig 2a) confirming that there is no positive susceptibility contribution from defects and oxygen vacancies of SnO_2 which has been cautioned to be a universal feature of nonmagnetic oxide nanoparticles [9].

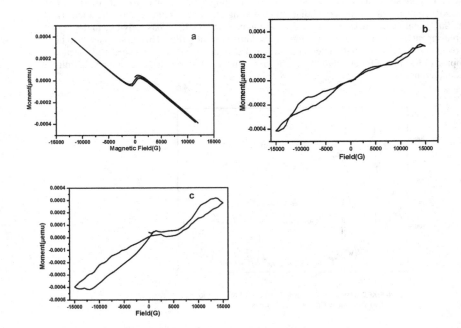

**Figure 2 : M – H curve of SnO_2:Mn films for Mn doping concentrations
a) 0 mol% Mn b) 1 mol% Mn c) 3 mol% Mn**

The diamagnetic behavior of undoped SnO_2 films suggests that the ferromagnetic signal in SnO_2: Mn films are solely due to the presence of Mn ions in SnO_2 host matrix. The M–H curves of SnO_2: Mn films(1 – 3 mol%) with higher Mn doping concentrations exihibit room temperature ferromagnetism. Since, the slope of the linear part of the curve increases with Mn concentration, we can attribute this variation to be dependent on Mn concentration. The enhancement of magnetization can be ascribed to the so-called F-centre exchange coupling, in which both oxygen vacancies and transition metal (TM) doping are involved [10]. The electron trapped in the oxygen vacancy occupies an orbital which overlaps the d shells of both Manganese neighbours. Based on Hund's rule and Pauli exclusion principle, spin orientations of the trapped electrons and the two neighbouring Mn ions should be parallel in the same direction, thus ferromagnetic ordering is achieved. This oxygen vacancy involving mechanism is further confirmed by the ferromagnetic behavior observed in Mn-doped SnO_2 samples[10] .

3.3 OPTICAL STUDIES

The optical properties of nanocrystalline semiconductors like SnO_2 have been studied extensively in recent years for translating their enhanced properties into practical applications. The optical transmission spectra of SnO_2: Mn thin films annealed in air at 500 °C are shown in Fig 3. The spectra show a high transmittance (>80 %) for the pure SnO_2 thin film and relatively lower transmittance for the films prepared by doping various Mn concentrations (1- 5 mol%). The higher transmittance observed in the pure SnO_2 film was attributed to less scattering effects, structural homogeneity and better crystallinity, whereas the lower transmittance in the SnO_2: Mn thin films can be due to a lesser crystallinity and increase in scattering centers (grain boundaries, defects, etc.)[12]. These interference patterns in the transmittance spectra are direct evidence for homogeneous and uniform films.

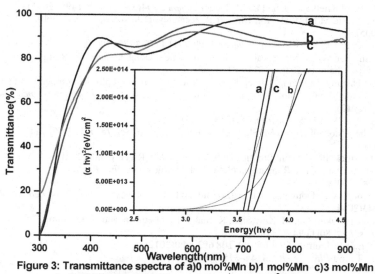

Figure 3: Transmittance spectra of a)0 mol%Mn b)1 mol%Mn c)3 mol%Mn in SnO_2:Mn thin films. (Inset: hυ versus $(αhυ)^2$ graph)

117

The optical band gap of SnO_2 : Mn films can be deduced from transmission measurements using Wood-Tauc relation [13]. E_{gap} values were obtained by extrapolating the linear portion of $(\alpha h v)^2$ versus (hv) plots, to intercept the photon energy axis (inset- Fig 3). In SnO_2 : Mn films, the plots of $(\alpha h v)^2$ as a function of energy(hv) for the films tend asymptotically towards a linear section, which shows that the investigated films have a direct optical band gap. The measured optical band gap increased from 3.55 to 3.71 eV with increase in the Mn concentration. The Eg of SnO_2 : Mn films were found to be inversely proportional to the crystallite size of the nanoparticles. This is due to the quantum confinement effect [14]. The change in band gap suggests that the size and shape of the nanoparticles influence the opto-electronic properties of the materials and can be tuned by doping.

3.4 CONCLUSION

This study systematically discussed the influence of Mn doping concentration on SnO_2 films prepared using sol–gel technique. Undoped and SnO_2 :Mn films were deposited by sol–gel dip–coating method onto corning glass substrates. The incorporation of Mn into the SnO_2 film caused a decrease in crystallite size. The optical band gap values increased from 3.55 to 3.71 eV with increasing Mn doping. An analysis of structural, magnetic, and optical properties of the SnO_2:Mn thin films indicated its excellent quality suitable for application as dilute magnetic semiconductor devices.

REFERENCES

[1] S. A. Wolf, D. D. Awschalom, R. A. Buhrman, J. M. Daughton, S. Von Molnar, M. L. Roukes, A. Y. Chtchelkanova, and D. M. Treger, *Science* 294[5546] **1488-1495** (2001).
[2] S.C. Erwin, L. Zu, M.I. Haftel, A.L. Efros, T.A. Kennedy, D.J. Norris, *Nature* 436[03832] **91-95** (2005).
[3] D.J. Norris, A.L. Efros, S.C. Erwin, *Science* 319[5871] **1776-1779** (2008).
[4] Feng Gu, Shu Fen Wang, Meng Kai Lu, Guang Jun Zhou, Dong Xu and Duo Rong Yuan J. Phys. Chem. B, 108[35] **8119-8123**(2004). .
[5] S. Sujatha Lekshmy, Georgi P. Daniel, K. Joy, "Microstructure and physical properties of sol gel derived SnO_2: Sb thin films for optoelectronic applications" *Applied Surface Science* 274 , 95– 100 (2013).
[6] Swanepoel R *J Phys E: Sci Instrum* 16:1214 (1983).
[7] M. M. Bagheri-Mohagheri and M. Shokooh-Saremi *Semicond. Sci. Tecnol.* 19 [6] **764 – 769** (2004).
[8] Cullity BD, Stock SR , 3rd edn. Prentice Hall, Upper Saddle River, p 388 (2001).
[9] Sundaresan A, Bhargavi R, Rangarajan N, Siddesh U, Rao CNR *Phys Rev B* 74:161306 (2006)
[10] J. M. D. Coey, A. P. Douvalis, C. B. Fitzgerald, and M. Venkatesan *Appl. Phys. Lett.* 84[8] **1332-1343**(2004).
[11] H. Kimura, T. Fukumura, M. Kawasaki, K. Inaba, T. Hasegawa, and H. Koinuma *Appl. Phys. Lett.* 80[1] **94-97** (2002)
[12] Ghodsi F.E., Mazloom J, Appl Phys A 108 **693–700**(2012).
[13]Wood DL, Tauc J *Phys Rev B* 5:**3144–3151**(1972).

[14]Valle GG, Hammer P, Pulcinelli SH, Santilli CV *J. Eur. Ceram. Soc* 24:**1009–1013** (2004).

Mater. Res. Soc. Symp. Proc. Vol. 1675 © 2014 Materials Research Society
DOI: 10.1557/opl.2014.874

Effect of Gd-substitution at Y-site on the structural, magnetic and dielectric properties of $Y_{1-x}Gd_xMnO_3$ (x=0, 0.05) nanoparticles

Samta Chauhan[1], Saurabh Kumar Srivastava[1], Amit Singh Rajput[1,2], Ramesh Chandra[1,*]
[1]Nanoscience Laboratory, Institute Instrumentation Centre, Indian Institute of Technology Roorkee, Roorkee-247667, India
[2]Centre of Nanotechnology, Indian Institute of Technology Roorkee, Roorkee-247667, India

* Corresponding author
Email: ramesfic@iitr.ernet.in

ABSTRACT

Effect of Gd substitution at Y-site on the structural and magnetic properties of $Y_{1-x}Gd_xMnO_3$ (x=0, 0.05) nanoparticles prepared by conventional solid state reaction method has been studied. The structural study using X-ray diffraction pattern indicates the hexagonal structure with $P6_3cm$ space group for all the samples. The average particle size for all the samples lies in the range of 30-40 nm as confirmed by X-ray diffraction and transmission electron microscopy analysis. The change in a and c lattice parameters confirm the substitution of Gd at Y-site. Magnetization versus temperature measurements show enhanced magnetic moment and an increase in Neel temperature with Gd-doping. Spin glass behavior is observed at low temperature in all the samples. Exchange bias effect has been observed at 5 K after field cooling the samples which is ascribed to the formation of antiferromagnetic-ferromagnetic (AFM-FM) core-shell structure of the nanoparticles. A significant improvement in the dielectric properties of Gd-doped samples has also been observed.

INTRODUCTION

Multiferroics are those which exhibit multiple ferroic properties: ferro-magnetic, ferro-electric and ferro-elastic (or at least two of them), combined in the same phase [1-6]. In these materials, the interaction between magnetic and electrical properties leads to control of magnetization by applied electric field and polarization by magnetic field [1]. Since the past few years, substantial research is going on the possibility to couple these ordering effects for applications in spintronics, memory devices, multiple-state logic devices, magneto-electric sensors etc [2,7,9,10]. The hexagonal manganites $RMnO_3$ with R=Ho-Lu and Y have been of particular interest for the detailed studies of the physics of multiferroics. $YMnO_3$ is among the hexagonal rare earth manganites with smaller R3+ ion radii, which belongs to a P63cm space group. The structure comprises of layers of MnO_5 bipyramids with one Mn^{3+} ion surrounded by three in-plane O ions and two apical O ions, separated by layer of Y^{3+}[8]. The in-plane O ions form 2D Mn-O-Mn super-exchange bond giving rise to an antiferromagnetic (AFM) state with $T_N \sim 70K$. The ferroelectric nature (with $T_C \sim 900K$) arises due to the tilting of MnO_5 polyhedra combined with displacement of Y^{3+} ions. In the series of investigation, numbers of research have been carried out to study the effect of doping in $YMnO_3$. Recent studies have focused on the phase transitions in high temperature limit [11] and magnetic phase diagrams of h-$YMnO_3$ [12]. With Cr doping o-YMO has also shown some multiferroic behavior but the transition temperature (T_N) was found to be very low (~25-30K) [13]. In case of $Ca_xY_{1-x}MnO_3$, the zero-field cooling (ZFC) curves shifted with higher temperature as Ca-doping x is increased [14]. The Exchange bias phenomenon has also been observed in multiferroic $Eu_{0.75}Y_{0.25}MnO_3$ which is found to exhibit weak ferromagnetism with cone spin configuration induced by external magnetic field below 30 K [15]. In this paper, we report the effect of Gd doping on structural, magnetic and dielectric properties of $YMnO_3$ nanoparticles. Gd possess similar ionic radius as Y and therefore maintained hexagonal structure of Gd doped $YMnO_3$ nanoparticles.

EXPERIMENTAL DETAILS

The polycrystalline samples of $YMnO_3$ and $Y_{0.95}Gd_{0.05}MnO_3$ nanoparticles were prepared using conventional solid state reaction method by mixing Y_2O_3 (99.99% purity), MnO_2 (99.99% purity) and Gd_2O_3 (99.99% purity) powders in a proper stoichiometric ratio. The mixed powders were ground for 12-13 hrs in an agate pestle mortar and then calcined for 3 hours at 1200°C. For dielectric measurements, the powders were pressed into pellets of 16 mm diameter and 1mm thickness using Hydraulic press. The pellets were again sintered in air at 1200°C for 24 hrs to ensure proper binding.

DISCUSSION

XRD Analysis

Figure 1 shows the X-Ray diffraction (XRD) pattern for $Y_{1-x}Gd_xMnO_3$ (x=0, 0.05) nanoparticles which confirm that all the samples have pure hexagonal crystal structure with $P6_3cm$ space group. The narrow and intense diffraction peaks show superior crystallization and grain growth of the synthesized samples at selected calcination temperature.

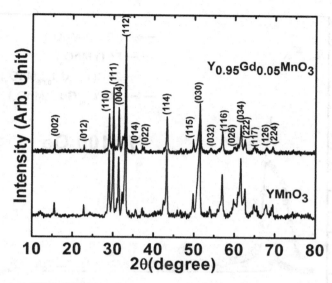

Figure 1. X-Ray Diffraction Patterns for $Y_{1-x}Gd_xMnO_3$ (x=0, 0.05) nanoparticles

The average particle sizes (D) calculated by Scherrer formula [16] were found to be 34 nm and 40 nm for $YMnO_3$ and $Y_{0.95}Gd_{0.05}MnO_3$ nanoparticles, respectively. The lattice constants obtained from the XRD pattern for $YMnO_3$ nanoparticles are a ~ 6.1581(2) Å and c ~11.3809(3) Å, while for $Y_{0.95}Gd_{0.05}MnO_3$ nanoparticles the lattice constants are a ~ 6.1464(2) Å and c ~11.3995(2) Å. It has been observed that the lattice constant a decreases and c increases with Gd doping at Y site in hexagonal $YMnO_3$ nanoparticles [17]. The crystallite size of hexagonal $YMnO_3$ nanoparticles increases with Gd doping, which may be due to higher ionic radius of Gd^{3+} as compared to Y^{3+} [17].

Magnetic Properties

Figure 2 shows the temperature dependence of *dc* magnetic susceptibility χ_{dc} of the samples under an applied field of 100 Oe with both ZFC and FC conditions in the temperature range 5-300 K. For both ZFC and FC measurements, the samples were cooled from 300 K to 5 K and the measurements were recorded during heating.

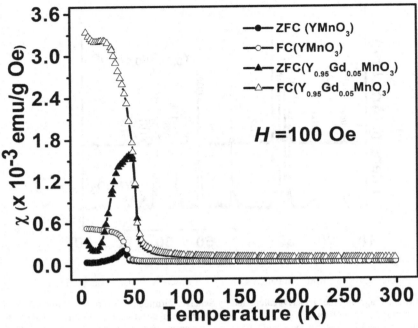

Figure 2. Temperature dependence of dc magnetic susceptibility for nanoparticles at H=100 Oe.

The magnetic moment in YMnO$_3$ nanoparticles is only due to Mn ions, because Y^{3+} is non-magnetic. The susceptibility increases with decrease in temperature because fluctuations of the magnetic moments decrease due to decrease in thermal energy on lowering the temperature and Mn moments start to order antiferromagnetically via superexchange interactions at ~77 K. This temperature is Neel temperature corresponding to antiferromagnetic ordering of Mn moments. On further decreasing the temperature, the YMnO$_3$ nanoparticles exhibit a sharp irreversibity between ZFC and FC susceptibility curves and ZFC curve shows a cusp at 41 K. This type of behaviour is a feature of spin glass systems [17] and temperature corresponding to the cusp of ZFC susceptibility is the freezing temperature (T$_f$) of spins. The irreversibility between ZFC and FC susceptibility curves is due to the different responses of the spins to applied (external) magnetic field. Above T$_f$, spins are randomly aligned due to geometric frustration. In zero field cooling through T$_f$, susceptibility depends upon local anisotropy, the effect of surface spins becomes more prominent due to large surface area of nanoparticles causing some uncompensated spins at the surface. Thus nanoparticles of YMnO$_3$ were considered to consist of antiferromagnetic core of compensated spins and ferromagnetic shell of uncompensated surface spins. Spin glass behavior was again observed in Y$_{0.95}$Gd$_{0.05}$MnO$_3$ nanoparticles but the spin freezing temperature was shifted to higher value ~47 K. A sudden increase in magnetization below 12 K was observed, which was not present in YMnO$_3$

nanoparticles. This increase may be due to ferromagnetic transition corresponding to the ordering of Gd ions in YMnO$_3$. Moreover, the magnetic susceptibility has increased by large amount via Gd doping in hexagonal YMnO$_3$. The large increase in magnetic moment is due to replacement of Y^{3+}, nonmagnetic ion, by Gd^{3+} ions having effective magnetic moment of 8.0 μB [14, 18]. The increase in spin freezing by Gd doping may also be due to more ferromagnetic component induced, which may increase the competition between ferromagnetic and antiferromagnetic component and hence increase T$_f$. The spin glass behavior is due to competition between ferromagnetic and antiferromagnetic components in the system and in Y$_{0.95}$Gd$_{0.05}$MnO$_3$ (x=0, 0.05) nanoparticles, this ferromagnetic component comes from their nanometre range particle size.

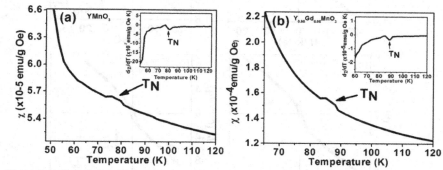

Figure 3. Enlarged view of the ZFC curves for (a) YMnO$_3$ and (b) Y$_{0.95}$Gd$_{0.05}$MnO$_3$ nanoparticles. Inset shows the dχ/dT Vs T curve.

The enlarged view of ZFC curve for YMnO$_3$ nanoparticles shown in figure 3(a) indicates an anomaly at ~ 77 K, which is further confirmed as a dip observed in dχ/dT plot (inset of figure 3(a)) thus confirming this temperature as Neel temperature. The plot of first order derivative of susceptibility with temperature was also used by Z. J. Huang *et al.* [19] to find the Neel temperature in YMnO$_3$. Figure 3(b) shows the enlarged view of ZFC curve for Y$_{0.95}$Gd$_{0.05}$MnO$_3$ nanoparticles. It clearly shows an anomaly at 88 K corresponding to the antiferromagnetic ordering of Mn moments. Thus, T$_N$ has increased with Gd doping in YMnO$_3$. The enhancement in T$_N$ was further supported by an anomaly at 88 K in dχ/dT plot for Y$_{0.95}$Gd$_{0.05}$MnO$_3$ nanoparticles as shown in the inset of figure 3(b). This change in T$_N$ is due to A- site cation ionic radius dependence of the magnetic properties in hexagonal manganites. Due to Gd substitution at Y-site, a change in Mn-Mn distance occur causing change in strength of exchange interactions between Mn ions which causes an increase in the Neel temperature. The increase in Neel temperature may also be due to increase in particle size [20].

To further explore the magnetic state of the nanoparticles, hysteresis *(M-H)* loops were measured at 5 K between ± 20 KOe. Figure 4 shows the M-H curves for the samples in ZFC and FC mode. The FC M-H loops were taken after cooling the samples from 300 K to 5 K in the presence of 1T magnetic field. It can be clearly seen that the magnetization varies linearly after a particular applied field and shows no saturation, which indicates the dominant role of AFM spins [21]. In addition, the FM behavior can also be seen in the hysteresis loops in the low field region.

The FM like behavior comes from the uncompensated surface spins of shell and the AFM linear contribution in the hysteresis loops is due to AFM core of compensated Mn-ion spins [21].

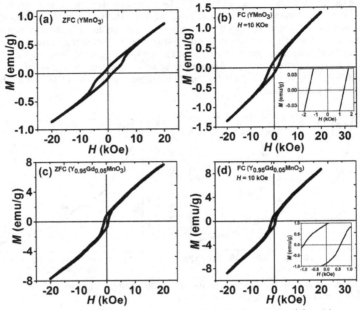

Figure 4. (a) ZFC, (b) FC M-H loops for YMnO$_3$ nanoparticles, (c) ZFC, (d) FC M-H loops for Y$_{0.95}$Gd$_{0.05}$MnO$_3$ nanoparticles

From figure 4(b) and (d) it is worth noting that there is a shift in the FC M-H loops along the negative applied field for both the samples, which indicates the presence of exchange bias (EB) in these samples. The exchange bias field is generally defined as $H_{eb}=|H_L+H_R|/2$, and coercivity is defined as $H_C=|H_L-H_R|/2$, where H_L and H_R are the left and right coercive fields, respectively. The shift in the M-H loops (H_{eb}) is found to be ~ 102 Oe for YMnO$_3$ and ~ 194 Oe for Y$_{0.95}$Gd$_{0.05}$MnO$_3$ nanoparticles. Inset of figure 4(b) and (d) shows the enlarged view of the M-H loops near low field region. The increase in the exchange bias in Y$_{0.95}$Gd$_{0.05}$MnO$_3$ nanoparticles compared to YMnO$_3$ nanoparticles is due to large magnetic moment induced by Gd ions [14,18].

Dielectric Properties

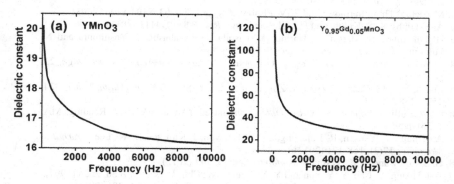

Figure 5. Frequency dependence of dielectric constant for $Y_{1-x}Gd_xMnO_3$ (x=0, 0.05) nanoparticles

Figure 5 shows the frequency dependence of dielectric constant for $Y_{1-x}Gd_xMnO_3$ (x=0, 0.05) samples. It shows that dielectric constant decreases with increase in frequency for both the samples. But the dielectric constant was increased by six times with the Gd doping. This may be due to change in the tilting of MnO_5 trigonal bipyramids.

CONCLUSION

In summary, we have studied the effect of Gd-substitution on the structural, magnetic and dielectric properties of h-$YMnO_3$. It was found that lattice constant a decreases while c increases with Gd-substitution. The T_N of the Gd-doped $YMnO_3$ nanoparticles increases to ~ 88 K which is ascribed to the change in Mn-Mn distance causing change in strength of exchange interactions between Mn ions. Exchange bias observed in the samples is attributed to the exchange interaction between the compensated AFM spins and uncompensated surface spins of the nanoparticles. Dielectric constant of the Gd-doped samples also increases due to change in tilting of MnO_5 trigonal bipyramids.

REFERENCES

1. Nicola A. Hill, *J. Phys. Chem. B* **104**, 6694-6709 (2000).
2. R. Ramesh and Nicola A. Spaldin, *Nat. Mater.* **6**, 21-29 (2007).
3. D.I. Khomskii, *J. Magn. Magn. Mater.* **306**, 1 – 8 (2006).
4. W. Eerenstein, N. D. Mathur & J. F. Scott, *Nature* **442**, 759-765 (2006).
5. N. A. Spaldin, S.W. Cheong, and R. Ramesh, *Phys. Today* **63**, 38-43 (2010).
6. C. N. R. Rao and C. R. Serrao, *J. Mater. Chem.* **17**, 4931-4938 (2007).
7. M. Bibes and A. Barthélémy, *Nat. Mater.* **7**, 425 - 426 (2008)
8. Bas B. Van Aken, Thomas T. M. Palstra, Alessio Filippetti and Nicola A. Spaldin, *Nat. Mater.* **3**, 164-170 (2004).

9. S. D. Sarma, *American Scientist* **89**, 516-523 (2001).
10. R. Ramesh, *Nat. Mater.* **9**, 380-381 (2010).
11. M. Fiebig, Th. Lottermoser and R. V. Pisarev, *J. Appl. Phys.* **93**, 8194-8196 (2003)
12. A. S. Gibbs, K. S. Knight and P. Lightfoot, *Phys. Rev. B* **83**, 094111 (2011).
13. H. Fukumura, S. Matsui, H. Harima, K. Kisoda, T. Takahashi, T. Yoshimura and N. Fujimura, *J. Phys.: Condens. Matter* **19**, 365239 (2007).
14. Y. Su, Z. Chen, Y. Li, D. Deng, S. Cao and J. Zhang, *J. Supercond. Novel Magn.* **23**, 501–506 (2010).
15. Li-Qin Yan, F. Wang, Y. Zhao, T. Zou, J. Shen and Y. Sun, *J. Magn. Magn. Mater.* **324**, 2579–2582 (2012).
16. B. D. Cullity, *Elements of X-Ray Diffraction*, 1st ed. (Addison-Wesley, Reading, MA, 1956), p. 99.
17. A. Ismail, W. Yansen, R. Rajagukguk, Y. M. Kwon, J. Kim and B. W. Lee, *Journal of Magnetics* **17(3)**, 168-171 (2012).
18. G. S. Lotey and N. K. Verma, *J. Nanopart Res.* **14**, 742 (2012).
19. Z. J. Huang, Y. Cao, Y. Y. Sun and Y. Y. Xue, C. W. Chu, *Phys. Rev. B* **56**, 2623 (1997).
20. Tai-Chun Han, Wei-Lun Hsu and Wei-Da Lee, *Nanoscale Res. Lett.* **6**, 201 (2011).
21. S. Chauhan, S. K. Srivastava, R. Chandra, *Appl. Phys. Lett.* **103**, 042416 (2013).

Mater. Res. Soc. Symp. Proc. Vol. 1675 © 2014 Materials Research Society
DOI: 10.1557/opl.2014.914

Influence of substrate temperature and post annealing on morphology and magnetic properties of pulsed laser deposited Fe$_{50}$-Ni$_{50}$ films.

Sally A. Ibrahim[1], Svitlana Fialkova[1], Kwadwo Mensah-Darkwa[1], Sergey Yarmolenko[1], and Dhananjay Kumar[1]
[1]North Carolina A&T State University,
1601 E. Market St. Greensboro, North Carolina, 27411, U.S.A.

ABSTRACT

As the need for smaller data storage devices in the market continues to grow, the study of new combinations of self-assembled magnetic nanoparticles/films is greatly needed. In this research, Fe$_{50}$-Ni$_{50}$ films were synthesized using a Pulsed Laser Deposition technique. The films were analyzed using scanning electron microscopy (SEM), atomic force microscopy (AFM) and physical properties measurement system (PPMS). Films were deposited from Fe-Ni alloy target (50%-50% composition), deposition was conducted in vacuum, at substrate temperatures varying from liquid nitrogen temperature -196°C to 600°C. The films were annealed in a vacuum chamber at 600°C for 1 hour. The study reveals that the substrate temperature has significant effect on the structure of the films and their magnetic properties. It was shown that additional thermal treatment improved the quality of films in terms of narrow grain size distribution. Magnetic properties were also found to improve significantly after post annealing process.

INTRODUCTION

Magnetic thin films have been utilized in many applications such as sensors, actuators and magnetic recording media (such as hard drives, CDs, DVDs and USB flash drives) [1, 2]. Researchers have been concentrating in increasing the storage capacity of these devices to compete with the miniaturization demands of the market. The FeNi alloys are considered good candidates for magnetic recording media because of its economic competence compared to FePt and FeCo which are noble and need complex processing. Previous studies on ZnO/Co multilayers deposited by pulsed laser deposition and annealed at elevated temperatures caused the thin films to transform from the amorphous phase to crystalline ZnO/Co phase. Formation of CoO and ZnCo$_2$O$_4$ which are antiferromagnetic along with the ferromagnetic metallic crystalline Co allowed FM-AFM coupling and therefore widening of the hysteresis loop [3]. The coercivity calculated from the hysteresis loop is an important parameter that indicates the resistance to demagnetization which is preferred in data storage devices. Soft magnetic alloys such as FeNi need manipulation to increase their coercivity[4]. In this study Fe$_{50}$Ni$_{50}$ thin films are fabricated by pulsed laser deposition at different substrate temperatures. The samples were then annealed in a vacuum chamber at 600°C for 1 hour. The as deposited and annealed thin films were characterized using SEM, AFM and PPMS to study the effect of post annealing on their magnetic properties.

EXPERIMENT

Fe$_{50}$Ni$_{50}$ alloy was prepared at University of Nebraska. Fe and Ni metal powders with 50% weight were mixed for 30 minutes in a sealed ball milling container. The powders were then melted using a vacuum arc melting system. The resulting melt was machined to produce the Fe$_{50}$Ni$_{50}$ target. In this study the Fe$_{50}$Ni$_{50}$ thin films were fabricated using (Excel Instruments) pulsed laser deposition (PLD) system on Silicon (100) substrates. The substrates were ultrasonically cleaned with acetone and methanol for 10 minutes each. The deposition chamber was flushed with argon gas at pressure 100 mTorr. The vacuum pressure at time of deposition was in the order of 10^{-7} Torr. Deposition parameters for the samples are shown in (Table 1). SEM was also used to verify AFM findings and to measure the composition of the thin films. Magnetic properties were extracted from the hysteresis loop obtained from the Quantum Design physical properties measurement system (PPMS). Magnetic measurements were conducted at temperatures of 10K, 100K, 200k and 300K.

Table1. PLD parameters for samples studied.

Sample ID.	Material	Vacuum Pressure Mbar	Substrate Temperature ° C	Laser Energy mJ	No. of pulses	Pulse rate Hz
32	Fe$_{50}$Ni$_{50}$	4.6 X 10^{-7}	25°	500	20,000	20
35	Fe$_{50}$Ni$_{50}$	3.4 X 10^{-7}	600°	500	20,000	20
37	Fe$_{50}$Ni$_{50}$	1.0 X 10^{-7}	Liquid N$_2$	500	20,000	20

AFM Topographic Measurements

The surface morphology was analyzed by using a NTEGRA Prima Modular scanning probe microscope from NT-MDT. The system was configured with the scanning by sample piezo-sensor (range 10x10 micron), and super sharp diamond-like carbon (DLC) tip with a typical spring constant of about 11.5 Nm^{-1} ,a resonance frequency around 255 kHz and a typical curvature radius 1 nm (K-Technano model no. NSG10-DLC) was used for scanning. The images were recorded in ambient conditions (25 °C and 20 % relative humidity) and in soft intermittent contact (semi-contact) mode. Tapping force was controlled by the ratio between set point amplitude (A$_{sp}$) and the free air amplitude (A$_0$). Several topographical images with areas 1.2x1.2 µm^2 and 0.6x0.6 µm^2 were recorded for each sample was, each scan area contains 512x512 (scan steps 2.35 nm and 1.1 nm). The scan rate was 0.5Hz. During acquisition of surface morphology, height images and amplitude error were recorded at the same time. Prior measurements the samples were cleaned in ultra-sonic bath with acetone (grade XXX) and methanol (grade XXX) and then blow-dried with nitrogen gas.

All off line image processing (flattening) and particles analyses were conducted at the software environment provided by the AFM manufacturer (Nova v1.1.0.1913 and IA-P9).

RESULTS AND DISCUSSION

Particles analysis

Particles analysis was performed as described above. Figure1 present the topographical images with areas 0.6x0.6 μm^2 of the samples as -deposited and after heat treatment. The particle size (height and diameter) for samples prior and after heat treatment are presented in table 1 and Figure 2.

The AFM images show that some of the films have shinny large particles that are likely to be metal droplets from the target known to be produced by the PLD [5]. Sample 37 deposited at liquid N2 temperature have a very fine smooth surface but after annealing its average surface roughness increased from 1.471nm to 3.789nm, which is a 160 % increase. This increase is attributed to the Ostwald ripening phenomena during heating. The same behavior is observed for sample 32 deposited at room temperature, its average surface roughness increased from 1.41nm to 7.25nm, which is a 414% increase. The results clearly shown that sample 35 deposited at 600°C produced stable uniformly sized and distributed particles.

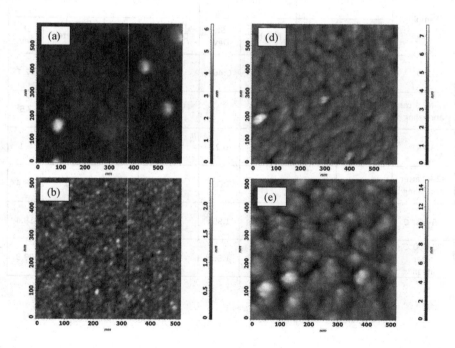

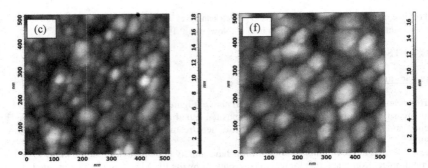

Figure 1. Samples topography a) deposited at liquid Nitrogen, b) RT, c) 600 C, and after annealing at 600°C for 1 hour: d), e) and f) respectively.

Table 2. Particles analysis

Sample	1μm		500nm		1μm		500nm	
	Particles Height (nm)	St. Dev.	Particles Height (nm)	St. Dev.	Particles Diameter (nm)	St. Dev.	Particles Diameter (nm)	St. Dev.
37-N2	2.382	1.514	1.471	0.465	49	17.1	38.6	12.1
37-N2 after annealing	6.553	0.438	3.789	0.39	22.5	10.3	23.8	7.5
32-RT	1.489	0.294	1.41	0.21	15.4	5.9	12.9	5.1
32-RT after annealing	17.275	9.201	7.25	1.065	51.4	23.6	46.2	18.4
35-600°C	18.676	3.152	9.046	1.951	72.5	19.8	59.7	18.4
35-600°C after annealing	18.071	2.842	9.979	1.866	82.3	18.5	71.6	23.3

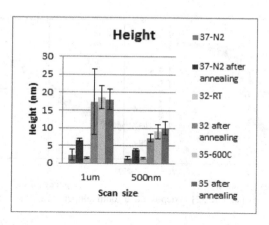

Figure 2 AFM measured values of particle height at 1 µm and 500 nm scan sizes

The as deposited film had a rough surface compared to the other two samples. Its average surface roughness only increased from 9.046 nm to 9.979 nm with an increase of 10%.

Magnetic Results:

The M-H curves of the as-deposited films and post annealing films obtained by PPMS are shown in Figures 3-5. The narrow loops and low coercivities are the known properties of soft magnetic materials. The coercivity of sample 35 deposited at 600°C is 44.76Oe almost as twice as that of sample 32 deposited at room temperature which is found to be 25.86Oe, this higher coercivity is due to the decreased number of defects on higher deposition temperatures and the thermal stability of the films also shown on the AFM results of the stable surface morphology after post annealing for this sample. Sample 35 also show formation of crystal structures. For high deposition temperature the substrate surface have high energy as well as the deposited particles which bounce back when it come into contact with the surface allowing for the formation of a very small numbers of seeds to adhere to substrates and eventually the particles mobility allow them to create large crystals at these seeds. Sample 37 deposited at liquid Nitrogen temperature is the most interesting since its hysteresis loop show superparamagnetic behavior. The nature of this deposition allows for the quick adhesion of very fine particles which explain the behavior of the hysteresis loop.

The coercivities are increased significantly post annealing. Coercivity of sample 32 post annealing increased from 25.86Oe to 173.54Oe at 10K which is a 571% increase. This is mostly due to the formation of crystal like structures on the sample. Sample 35 post annealing coercivity increased from 44.76Oe to 208.79Oe at 10K which is an increase of 4 %; this assumed to be caused by a sort of adjustment of the crystal orientation. Sample 37 as deposited magnet behavior changed from suparamagnetic to ferromagnetic exhibiting a hard magnetic material wide hysteresis loop.

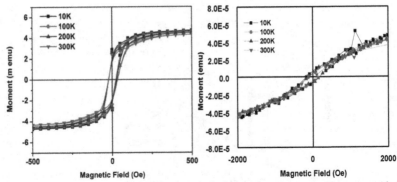

Figure 3. Hysteresis loops [sample 32] a) deposited at room temperature, b) post annealed respectively.

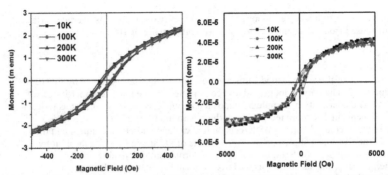

Figure 4. Hysteresis loops [sample 35] a) deposited at 600˚C, b) post annealed respectively.

This change indicates formation of an $L1_0$ structure of the FeNi. The $L1_0$ FeNi has high coercivities in the range of 500Oe to 3KOe [6] because of its high uniaxial magnetic anisotropy and good thermal stability of magnetization. Figure 6 shows the significant increase in the coercivities of the as deposited films and post annealed films.

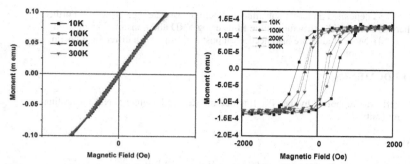

Figure 5. Hysteresis loops [sample 37] a) deposited at Liquid Nitrogen, b) post annealed respectively.

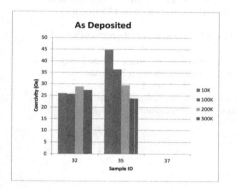
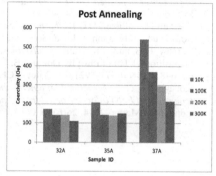

Figure 6. Coercivity comparison of the as deposited and post annealed films at measurement temperatures of 10K, 100K, 200K and 300K.

CONCLUSIONS

To study the post annealing influence on magnetic properties, FeNi thin films were deposited at room temperature, 600°C, and liquid Nitrogen temperature, and then annealed at 600°C for 1 hour in a vacuumed chamber. Surface roughness obtained by AFM show that annealing increased the surface roughness of samples deposited at RT, and Liquid N_2 .Results from the hysteresis loops show that post annealing had a significant effect on magnetic properties. The film deposited at Liquid Nitrogen temperature exhibited perfect superparamagnetic behavior. Post annealing changed this behavior to a hard ferromagnetic behavior. XRD results were insuffiecient because of size limitations of sample size needed for magnetic measurements. Also post annealed films deposited on larger substrate will help investigate their structure. Further studies are needed to figure the structure of the films using XRD and transmission electron microscopy (TEM). More studies are also needed on post annealing with different annealing temperatures and durations.

ACKNOWLEDGMENTS

This work was supported by the NSF Engineering Research Center at North Carolina A&T State University.

REFERENCES

1. Brauer, J.R., *Magnetic Actuators and Sensors*. Wiley - IEEE Press.
2. Frey, N.A. and S. Sun, *Magnetic Nanoparticle for Information Storage Applications.* Inorganic Materials, 2010: p. 33-68.
3. Wang, Z., et al., *The influence of layer thickness and post annealing on magnetism of pulsed laser deposited ZnO/Co multilayers.* Journal of Magnetism and Magnetic Materials, 2013. **345**(0): p. 41-47.
4. William D. Callister, D.G.R., *Materials Science and Engineering: An Introduction.* 8th Edition ed. 2009: John Wiley & Sons, Inc.
5. Chrisey, D.B. and G.K. Hubler, *Pulsed laser deposition of thin films.* Pulsed Laser Deposition of Thin Films, by Douglas B. Chrisey (Editor), Graham K. Hubler (Editor), pp. 648. ISBN 0-471-59218-8. Wiley-VCH, May 2003., 2003. **1**.
6. Kotsugi, M., et al., *Structural, magnetic and electronic state characterization of L1 0-type ordered FeNi alloy extracted from a natural meteorite.* Journal of Physics: Condensed Matter, 2014. **26**(6): p. 064206.

Oxide Thin Films and Nanostructures

Mater. Res. Soc. Symp. Proc. Vol. 1675 © 2014 Materials Research Society
DOI: 10.1557/opl.2014.864

Photoactivated Metal-Oxide Gas Sensing Nanomesh by Using Nanosphere Lithography

Yu-Hsuan Ho[1,2], Tsu-Hung Lin[3], Yi-Wen Chen[3], Wei-Cheng Tian[1],
Pei-Kuen Wei[2], and Horn-Jiunn Sheen*[3]
[1]Graduate Institute of Electronics Engineering, National Taiwan University, Taipei, Taiwan
[2]Research Center for Applied Sciences, Academia Sinica, Taipei 115, Taiwan
[3]Institute of Applied Mechanics, National Taiwan University, Taipei, Taiwan

ABSTRACT

A photoactivated ZnO nanomesh with precisely controlled dimensions and geometries is fabricated by using nanosphere lithography process. The nanomesh structures effectively increase the surface-to-volume ratio to improve the sensing response under the same testing gas. And the periodical nanostructures also increase the effective light path and lead to more efficient light activation for gas sensing. With the increase of the photoinduced oxygen ions by UV illumination, a distinguished sensing response is observed at room temperature. In the optimized case, the sensing response ($\Delta R/R_0$) of the ZnO nanomesh at the butanol concentration of 500 ppm is 97.5%, which is 4.54 times higher than the unpatterned one.

INTRODUCTION

Gas sensors have been a focus of research in recent years for various applications, such as breath tests, environmental monitoring, indoor air quality, workplace health and safety, and homeland security. There have been numerous attempts to develop sensing devices with high sensitivity, stability, and rapid response[1-4]. There are four main basic concepts of gas sensors that have so far been reported including chemoresistive sensor[5-8], capacitive sensor[9, 10], micromachined cantilever[11, 12] and microcalorimeter platform[13]. Among various types of sensing technologies, the chemoresistive gas sensor is one of the most promising methods due to its simple operation and high sensitivity.

To provide sufficient reaction energy for oxidization, metal oxide gas detectors are typically operated above 100-200 °C to achieve a highly sensitive operation with a fast sensing response[14]. However, the necessity of high operation temperatures for conventional metal oxide gas-based detectors affects the potential usage of these devices, such as in those that require low energy consumption. Recently, several approaches have been proposed to reduce the operation temperature of detectors, such as using doped metal in metal oxide materials[15], improving thermal isolation using MEMS technologies[16], alternative nanosensing materials [17], and the incorporation of UV illumination during detection[14, 18-20]. Among these techniques, the application of UV illumination on metal oxide detectors is one of the most promising methods for achieving room temperature gas sensing. With this illumination, in which near-UV radiation is used in the heterogeneous photocatalysis[21, 22], the metal oxide-based detector can significantly increase the bonding sites of the sensing materials at room temperature and increase the sensitivity of the detector. Some photocatalytic metal oxide, such as titanium oxide (TiO_2) or zinc oxide (ZnO) with an energy gap of approximately 3 eV is an effective photoactivated sensing material[23]. When photocatalytic metal oxide is under UV illumination, the ambient oxygen molecules are adsorbed to the its surface and are ionized to form O_2^-, O^-, or O^{2-} by capturing photoactivated free electrons from the conduction band of the metal oxide layer. The increase of these photoinduced oxygen ions, which easily react with reducing gases[19], is used for the reaction with target gases, resulting in significantly increased sensitivity. Therefore, UV illumination can effectively increase sensing performance and achieve room temperature gas sensing. Several research groups have reported on UV-activated metal oxide gas sensors at room temperature[14, 24, 25]. In this paper, we fabricated a ZnO nanomesh as the gas sensing layer using nanosphere lithography. The improved photoactivated sensing responses of the ZnO nanomesh was achieved due to the availability of the ultrafine nanostructures and the increased effective light path.

EXPERIMENT

The schematic diagram of fabrication process is shown in Figure 1. At first, commercial polystyrene (PS) nanospheres were dispersed in a solution of methanol and surfactant Triton X-100 and spun-coated onto a clean flat glass substrate. After the mixing process with sonic vibration, the prepared nanosphere solution had been spun-coated on the clean glass substrate. By controlling the concentration of nanosphere solution and rotation speed of the spin coater, the nanospheres were self-assembled to form a monolayer with closed-packed arrangement. After the closed-packed nanosphere monolayer is self-assembled, the O_2 RIE etching is applied to decrease the diameter of nanospheres. If the etching power is too low, it leads to the nonuniform pattern as shown in Figure 2-10(a). The gaps between spheres are not equally wide maybe due to the non-uniform O_2 ions. But a too high etching power would also result in the sphere's shape deformation. The proper etching power can effectively reduce the nanosphere size and maintain the sphere shape.

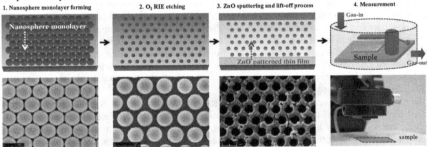

Figure 1. Nanosphere lithography for the photoactivated ZnO Gas Sensing Nanomesh

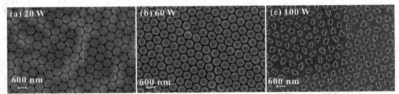

Figure 2. SEM images of nanosphere monolayer under different etching power

140

DISCUSSION

The period of the hexagonal array was 500 nm which is determined by the diameter of PS spheres. Then oxygen reactive-ion etching (RIE) is used to etch the edge of the PS nanospheres and reduce their size in order to adjust the fill-factor, which is defined by the ratio of the area of nanohole array divided by the total area. The corresponding SEM images of nanospheres under different etching condition are shown in Figure 3. The diameter of nanospheres is usually negative correlated to the etching time as shown in Figure 3(d). Subsequently, electron beam evaporation is used to deposit ZnO 200 nm on the substrate to fill the gap of nanospheres. The sample was then put into the dichloromethane (CH_2Cl_2) solution in an ultrasonic cleaner. After lift-off process, the PS nanospheres were removed and the ZnO remained on the substrate forming a periodic silver nanomesh as shown in Figure 4. The patterned thin film was further treated with an annealing process at 300°C for 1 hour by using high temperature furnaces. The annealing treatment is crucial to the reduction the background defects and decrease of the resistivity of the nanomesh.

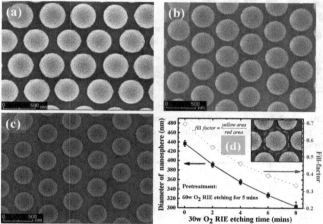

Figure 3. SEM images of close-packed PS nanospheres with different etching conditions: (a) RIE power of 60W for 3.5 minutes; (b) RIE power of 60W for 3.5 minutes and 30W for 4 minutes; (c) RIE power of 60W for 3.5 minutes and 30W for 8 minutes. (d) Size of PS nanosphere as a function of etching time

Gas sensors are calibrated and tested using a test system where test sample gas of defined mixture composition is generated from vapor delivery sub-system. The vapor delivery sub-system is utilized to convert the liquid phase test sample into a vapor phase prior to inject into a test chamber. The carrier gas is pumped through an air compressor and flowed through the molecular sieve, activated carbon and high efficiency particulate air (HEPA) filter sequentially before entering the bubbler. The molecule sieve and activated carbon can help to filter out the water vapor and organic impurities respectively. And HEPA is proposed to remove fine particulate matters. The clean carrier gas is then bubbled through the liquid. The rate of carrier gas flow into the reservoir is set with the mass flow controller (MFC). The carrier gas brings the vapor of the test sample into the mixing tube at a constant flow rate. In the mixing tube, the sample is mixed with the clean air to form various concentrations for device characterization.

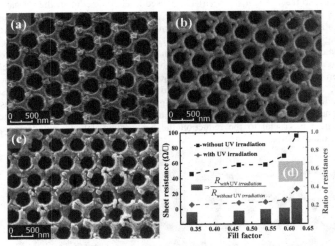

Figure 4. SEM images of close-packed PS nanospheres with different etching conditions: (a) RIE power of 60W for 3.5 minutes; (b) RIE power of 60W for 3.5 minutes and 30W for 4 minutes; (c) RIE power of 60W for 3.5 minutes and 30W for 8 minutes. (d) Size of PS nanosphere as a function of etching time

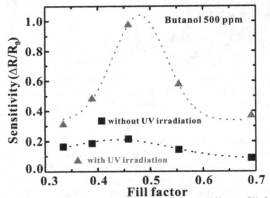

Figure 5. Sensitivity of the ZnO nanomesh with different fill-factor

The fill-factor directly involves the electricity of the nanomesh. The sheet resistance of the nanomesh with different fill-factor was shown in Figure 4(d). When the fill-factor increased, the finer nanostructures resulted in a rapid increase of the sheet resistance. This may be due to some inevitable breakages of the nanomesh in the lift-off process. There is a tradeoff between fill-factor and sensing performance. The depletion layer of the ZnO nanomesh with high fill-factor under gas exposure dominates its electrical properties, so the fine nanostructures are generally thought to be beneficial for the sensing response (ΔR). But the increased resistance would reduce the sensitivity ($\Delta R/R_0$). After being exposed to UV irradiation, the sheet resistance was lowered down by about 5 times, and the sensitivity had been greatly improved. The results of the sensitivity for butanol gas were shown in Figure 5. The optimized condition for the ZnO nanomesh is at the fill-factor 0.458.

CONCLUSIONS

We had successfully demonstrated a ZnO nanomesh with precisely controlled dimensions and geometries by using nanosphere lithography. With the UV activation, the gas sensing responses of ZnO nanomesh were observed at room temperature. In the optimized case, the sensing response ($\Delta R/R_0$) of the ZnO nanomesh at the butanol concentration of 500 ppm is 97.5%, which is 4.54 times higher than the unpatterned one.

REFERENCES

1. M.J. Deen, M.H. Kazemeini and S. Holdcroft: Contact effects and extraction of intrinsic parameters in poly(3-alkylthiophene) thin film field-effect transistors. *Journal of Applied Physics* **103**, 124509 (2008).
2. M.M. Arafat, B. Dinan, S.A. Akbar and A.S.M.A. Haseeb: Gas Sensors Based on One Dimensional Nanostructured Metal-Oxides: A Review. *Sensors* **12**, 7207 (2012).
3. C. Wang, L. Yin, L. Zhang, D. Xiang and R. Gao: Metal Oxide Gas Sensors: Sensitivity and Influencing Factors. *Sensors* **10**, 2088 (2010).
4. M.E. Franke, T.J. Koplin and U. Simon: Metal and metal oxide nanoparticles in chemiresistors: does the nanoscale matter? *Small* **2**, 36 (2006).
5. J.S. Suehle, R.E. Cavicchi, M. Gaitan and S. Semancik: Tin oxide gas sensor fabricated using CMOS micro-hotplates and in-situ processing. *Electron Device Letters, IEEE* **14**, 118 (1993).
6. W.-C. Tian, Y.-H. Ho, C.-H. Chen and C.-Y. Kuo: Sensing Performance of Precisely Ordered TiO2 Nanowire Gas Sensors Fabricated by Electron-Beam Lithography. *Sensors* **13**, 865 (2013).
7. Q. Fang, D.G. Chetwynd, J.A. Covington, C.S. Toh and J.W. Gardner: Micro-gas-sensor with conducting polymers. *Sensors and Actuators B: Chemical* **84**, 66 (2002).
8. C.-L. Li, Y.-F. Chen, M.-H. Liu and C.-J. Lu: Utilizing diversified properties of monolayer protected gold nano-clusters to construct a hybrid sensor array for organic vapor detection. *Sensors and Actuators B: Chemical* **169**, 349 (2012).
9. T.J. Plum, V. Saxena and J.R. Jessing: Design of a MEMS capacitive chemical sensor based on polymer swelling, in Microelectronics and Electron Devices, 2006. WMED '06. 2006 IEEE Workshop on (2006), pp. 2 pp.
10. H. Baltes, D. Lange and A. Koll: The electronic nose in Lilliput. *Spectrum, IEEE* **35**, 35 (1998).

11.   H.P. Lang, R. Berger, F. Battiston, J.P. Ramseyer, E. Meyer, C. Andreoli, J. Brugger, P. Vettiger, M. Despont, T. Mezzacasa, L. Scandella, H.J. Güntherodt, C. Gerber and J.K. Gimzewski: A chemical sensor based on a micromechanical cantilever array for the identification of gases and vapors. *Appl Phys A* **66**, S61 (1998).

12.   M. Maute, S. Raible, F.E. Prins, D.P. Kern, H. Ulmer, U. Weimar and W. Göpel: Detection of volatile organic compounds (VOCs) with polymer-coated cantilevers. *Sensors and Actuators B: Chemical* **58**, 505 (1999).

13.   J. Lerchner, J. Seidel, G. Wolf and E. Weber: Calorimetric detection of organic vapours using inclusion reactions with organic coating materials. *Sensors and Actuators B: Chemical* **32**, 71 (1996).

14.   E. Comini, A. Cristalli, G. Faglia and G. Sberveglieri: Light enhanced gas sensing properties of indium oxide and tin dioxide sensors. *Sensors and Actuators B: Chemical* **65**, 260 (2000).

15.   L.C. Tien, P.W. Sadik, D.P. Norton, L.F. Voss, S.J. Pearton, H.T. Wang, B.S. Kang, F. Ren, J. Jun and J. Lin: Hydrogen sensing at room temperature with Pt-coated ZnO thin films and nanorods. *Applied Physics Letters* **87**, 22106 (2005).

16.   S.P. Arnold, S.M. Prokes, F.K. Perkins and M.E. Zaghloul: Design and performance of a simple, room-temperature Ga2O3 nanowire gas sensor. *Applied Physics Letters* **95** (2009).

17.   C. Young-Jin, H. In-Sung, P. Jae-Gwan, C. Kyoung Jin, P. Jae-Hwan and L. Jong-Heun: Novel fabrication of an SnO 2 nanowire gas sensor with high sensitivity. *Nanotechnology* **19**, 095508 (2008).

18.   K. Anothainart, M. Burgmair, A. Karthigeyan, M. Zimmer and I. Eisele: Light enhanced NO2 gas sensing with tin oxide at room temperature: conductance and work function measurements. *Sensors and Actuators B: Chemical* **93**, 580 (2003).

19.   S.-W. Fan, A.K. Srivastava and V.P. Dravid: UV-activated room-temperature gas sensing mechanism of polycrystalline ZnO. *Applied Physics Letters* **95**, 142106 (2009).

20.   W.-C. Tian, Y.-H. Ho and C.-H. Chou: Photoactivated TiO2 Gas Chromatograph Detector for Diverse Chemical Compounds Sensing at Room Temperature. *Sensors Journal, IEEE* **13**, 1725 (2013).

21.   Y. Shapira, S.M. Cox and D. Lichtman: Photodesorption from powdered zinc oxide. *Surface Science* **50**, 503 (1975).

22.   Y. Shapira, R.B. McQuistan and D. Lichtman: Relationship between photodesorption and surface conductivity in ZnO. *Physical Review B* **15**, 2163 (1977).

23.   E. Comini, G. Faglia and G. Sberveglieri: Solid state gas sensing, (Springer2009).

24.   J. Saura: Gas-sensing properties of SnO2 pyrolytic films subjected to ultrviolet radiation. *Sensors and Actuators B: Chemical* **17**, 211 (1994).

25.   E. Comini, G. Faglia and G. Sberveglieri: UV light activation of tin oxide thin films for NO2 sensing at low temperatures. *Sensors and Actuators B: Chemical* **78**, 73 (2001).

Mater. Res. Soc. Symp. Proc. Vol. 1675 © 2014 Materials Research Society
DOI: 10.1557/opl.2014.877

Synthesis and Characterization of NaNbO₃ Mesostructure by a Microwave-Assisted Hydrothermal Method

Guilhermina F. Teixeira[1], Maria Ap. Zaghete[1], José A. Varela[1], Elson Longo[1].
[1]Chemistry Institute - UNESP, 55, Professor Francisco Degni Street, Araraquara, Brazil

ABSTRACT

In the present work, we report the synthesis and characterization of NaNbO₃ particles obtained by microwave-assisted hydrothermal method from Nb₂O₅ and NaOH. The synthesis was made at different periods at 180 °C and 300W. The crystallization of NaNbO₃ structures produced Na₂Nb₂O₆.H₂O in the intermediate phase with fiber-like morphology, and this is associated with the synthesis time. Pure orthorhombic NaNbO₃ with cube-like morphology originates after synthesizing for 240 minutes. To verify the remnant polarization of particles, films were obtained by electrophoresis process and sintered at 800°C for 10 minutes in a microwave furnace. The films characterization indicated that films of niobate with fiber-like morphology present remaining polarization, and the morphology of cubes did not show remaining polarization. Considering these results, it can be concluded that the morphology implemented ferroelectric property of NaNbO₃.

INTRODUCTION

Perovskite compounds (ABO₃) attracted great scientific interest for having electrical and optical properties, making them applicable in several technology fields. Lead base materials, such as PZT, fall into this class because they present efficient anisotropic properties, like piezoelectric, ferroelectric and optical properties [1]. The efficiency of these materials is related to the morphology. When they present one-dimensional morphology, anisotropic properties can be improved [2].

These materials are easily obtained by chemical routes including the microwave hydrothermal synthesis which is a variation of the conventional hydrothermal method. This method is the starting point for the preparation of several perovskite materials that can be run at temperatures and times below those of conventional synthesis method. For example, Pechini method allows structural and morphological control of the material by changing the synthesis parameters such as, pH, reagent concentration, temperature and synthesis time periods [3].

Despite good PZT properties, however, materials like PZT are prone to suffer environmental damage and eventually may cause need for substitution of these materials. Alkali niobates, such sodium niobate (NaNbO₃), are promising materials for substituting materials like PZT [4,5]. Currently, the conventional hydrothermal synthesis of these materials has been extensively explored, but there are few studies reporting the NaNbO₃ by microwave hydrothermal synthesis.

Thus, the purpose of the present work is to investigate the properties of NaNbO₃ obtained by microwave hydrothermal synthesis.

EXPERIMENTAL DETAILS

The microwave hydrothermal synthesis of NaNbO₃ structures was performed beginning with NaOH (Quemis) and Nb₂O₅ (Alfa Aesar 99%). The reaction was carried out in a Teflon vessel model XP-1500 (CEM Corp.), in a MARS-5(CEM Corp.) microwave. The preparation conditions of the suspensions precursor are represented in Table I. This resultant suspension was then transferred into Teflon vessels and placed inside a microwave furnace. The synthesis was carried out at 180 °C and 300 W for different time periods.

Table I: Microwave hydrothermal synthesis conditions of $NaNbO_3$ particles.

Sample	Experimental Conditions					
	H_2O (mL)	Nb_2O_5 (g)	NaOH (mol.L^{-1})	Synthesis time (min.)	Temperature (°C)	Power (W)
NN1	30	0.696	7.5	60	180	300
NN2	30	0.696	7.5	120	180	300
NN3	30	0.696	7.5	180	180	300
NN4	30	0.696	7.5	240	180	300

The powders obtained were thoroughly washed with water by centrifugation and finally dried at room temperature. After that the NN1 product was calcined at 550°C/ 240 minutes. The obtained particles were characterized by X-ray powder diffraction (XRD) using a Rigaku-DMax/2500PC, with Cu-Kα radiation (λ = 1.5406 Å) in the 2θ range from 20 to 75° with 0.20/min. Raman spectra were collected (Bruker RFS-100/S Raman spectrometer with Fourier transform). A 1064 nm YAG laser was used as the excitation source, and its power was kept at 150 mW. The morphology of as-prepared samples was observed using a high resolution field emission gun scanning electron microscopy FE-SEM (GERMA JEOL JSM 7500F Field Emission Scanning Electron Microscopy). All measurements were performed at room temperature.

The samples of $NaNbO_3$ are prepared as films using the electrophoretic deposition (EDP) technique to characterize its ferroelectric properties. The films were deposited on $Pt/TiO_2/SiO_2/Si$ substrate using a voltage of 2 KV for a time period of 10 minutes to a stable ethylic suspension containing 0.00675 g of $NaNbO_3$ particles for promoting a uniform and rapid deposition. Following deposition, the films were heated using microwave radiation at 800°C for 10 minutes. The film preparation started with NN1 product after thermal treatment and NN4 product was obtained. They were subsequently characterized by FE-SEM and Ferroelectric hysteresis measurement at room temperature performed in a 400 VOLT Amplifier Radiant Technologies, Inc.

DISCUSSIONS

Figure 1 shows the diffraction patterns of the powders obtained by microwave hydrothesis. According to results, it was observed that the $NaNbO_3$ crystalline phase is favored by increase of synthesis time. For NN 1 and NN2 products, we noted that they were composed mostly of $Na_2Nb_2O_6.H_2O$. After the thermal treatment, NN1 product was converted to pure orthorhombic $NaNbO_3$ crystal structure according to the Joint Committee on Powder Diffraction Standards database (JCPDS) n° 33.1270. The peaks associated with $NaNbO_3$ in the form of NN3 became more apparent, while NN4 product obtained a pure $NaNbO_3$ (JCPDS n° 33. 1270). Song et al. also reported similar results of favoring the formation of $NaNbO_3$ crystalline phase with increase of synthesis time [6]. Paula et al obtained cubic particles of $NaNbO_3$ free of secondary phase at a temperature of 200° C at 60-minutes of synthesis time [7].

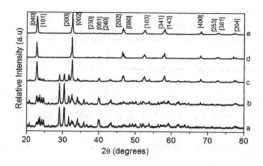

Figure 1: XRD patterns of particles obtained by microwave hydrothermal synthesis: (a) NN2, (b) NN3, (c) NN4, (d) NN5 and (e) NN1 after thermal treatment.

Figure 2 and Figure 3 represent the Raman spectra of the products obtained and provide structural information of the particles at short range. The images show the difference between the spectra obtained with the syntheses conditions. Spectra of products obtained at times lower than NN4 product show four bands located in 210, 457, 603 and 880 cm⁻¹ (Figure 2), while for NN4 product and NN1 product after thermal treatment 10 bands are observed (Figure 3).

The NaNbO₃ spectra (Figure 3) contain the sum of Na₂Nb₂O₆.H₂O and of NaNbO₃ vibrational modes. The identification of the Raman vibration modes can be considered a correlation between the point group symmetries of each atom in the unit cell, and the lattice space group symmetry (P_{bcm} forNaNbO₃ and $C_{2/c}$ for Na₂Nb₂O₆.H₂O) which generates the vibrational modes of each atom separately. Moreover, the internal modes results from the NbO₆ octahedron and the translation modes of the remaining cations' motion also are considered.

In the spectra, the modes found around 100 and 160 cm⁻¹ refer to NbO₆⁻ librational modes represented by $3A_g + 3A_u + 3B_{1g} + 3B_{1u} + 3B_{2g} + 3B_{2u} + 3B_{3g} + 3B_{3g}$, in an C_1 active site in a D_{2h} unit cell, being 12 Raman active vibrational modes [8].

The bands found below 500 cm⁻¹ are associated with the Nb-O-Nb vibration modes $F_{1u}(v_4)$, $F_{2g}(v_5)$ e $F_{2u}(v_6)$ and refer to local symmetry. The band in 180 cm⁻¹ is associated with F_{2u} mode. The existing modes in 219-276 cm⁻¹ refer to the F_{2g} mode, and the band present in 429 cm⁻¹ is associate with the F_{1u} antisymmetric mode. The presence of F_{1u} and F_{2u} modes is due to the multiple interaction in the unit cells, resulting in relaxation of selection rules, because the F_{1u} mode are active only in the infrared and F_{2u} mode is inactive. The bands in the region of 122 and 144 cm⁻¹ represent the Na⁺ translational modes with the NbO₆ octahedra. Wavelengths larger than 500 cm⁻¹ are related to $A_{1g}(v_1)$, $E_g(v2)$ e $F_{1u}(v3)$ stretches. The band found in 879 cm⁻¹ refers to F_{1u} mode Nb=O antisymmetric stretch, and the mode A_{1g} in 600 cm⁻¹ corresponds to the shortest bond between the niobium and oxygen in the NbO₆ octahedra of Na₂Nb₂O₆.H₂O, the E_g stretch mode appears in 560 cm⁻¹. The band existing at wavelength 663 cm⁻¹ refers to NbO₆. octahedra bonds. The smaller the length of the Nb-O bond the higher the spectrum vibration frequency [7]. Similar vibrations were obtained by analysis of other materials containing the NbO₆, as is the case for the KNbO₃, LiNbO₃, Fe/NaNbO₃ Ni/NaNbO₃, Co/NaNbO₃ e Ag/NaNbO₃ [9].

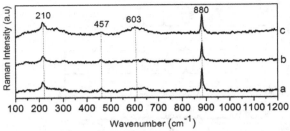

Figure 2: Raman spectra of microwave hydrothermal products: (a) NN1 before thermal treatment, (b) NN2 and (c) NN3.

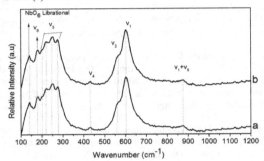

Figure 3: Raman spectra of microwave hydrothermal products: (a) NN1after thermal treatment, (b) NN4.

Figure 3 show the photomicrographs obtained using field emission scanning electron microscopy (FE-SEM). Different morphologies are observed for synthesis products. Note that the NN1 and NN2 samples are composed of $Na_2Nb_2O_6.H_2O$ structures with fiber-like shape (Figure 3 a-b), the NN3 sample presents a mix of fiber-like and plate-like structures (Figure 3c), and the NN4 shows cube-like structures (Figure 3d). These results corroborate with XRD diffratograms results (Figure1) because it is possible to observe that by increasing the synthesis time, the fiber-like morphology corresponding to the $Na_2Nb_2O_6.H_2O$ structure decreases, and after thermal treatment the NN1 product still presents fiber-like shape (Figure 3e) but with orthorhombic $NaNbO_3$ crystal structure (Figure 1). These results indicate that $Na_2Nb_2O_6.H_2O$ particles can be used as precursors to fabricate $NaNbO_3$ fibers without changing the fiber-like morphology by thermal treatment [10].

The formation of $NaNbO_3$ cubic particles starts with the niobium dissolution and the metastable phase ($Na_2Nb_2O_6.H_2O$). These structures can to present fiber-like or wheel-like structures compounded by NbO_6 octahedral, and the cube-like $NaNbO_3$ structures are formed by fibers dissolved under hydrothermal conditions [10, 11]

The dissolution of the particles is the consequence of the coarsening process Ostwald ripening. This process is characterized by growth of a determined particle through the dissolution of smaller ones and their subsequent deposition over the surface of greater particles. This process occurs spontaneously, in an attempt to decrease the total surface energy and occurred just after nucleation [12].

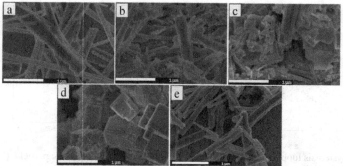

Figure 4: Images obtained by FE-SEM: (a) NN1 before thermal treatment, (b) NN2, (c) NN3, (d) NN4 and (e) NN1 after thermal treatment.

The images of NaNbO$_3$ films' surface are represented in Figure 4. From the images, it can be observed that the films exhibit a homogeneous yet porous covering. This was due to the sintering time which was not long enough for the densification of the material.

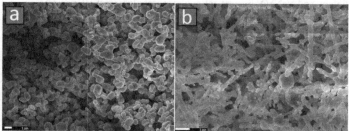

Figure 5: Images obtained by FE-SEM: (a) film obtained with cubic-like particle, (b) film obtained with fiber-like particles.

Figure 6 shows the polarization versus electric field applied for NaNbO$_3$ films with different morphologies. The NaNbO$_3$ films obtained from fibers-like particles (Figure 6b) exhibit remaining polarization (Pr) around 0.10μC/cm^2 and the film with cubic-like particles does not exhibit remnant polarization (Figure 6a). This indicates that the one-dimensional particles morphology improves the ferroelectric characteristic since this is an anisotropic property.

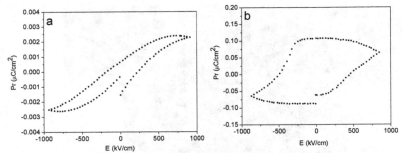

Figure 6: PE hysteresis loop of NaNbO₃ films: (a) film obtained with cubic-like particle, (b) film obtained with fiber-like particles.

CONCLUSION

In summary, $Na_2Nb_2O_6.H_2O$ and $NaNbO_3$ particles were obtained by microwave hydrothermal synthesis. The product formation is related to synthesis time, since the orthorhombic $NaNbO_3$ pure crystalline structure is favored by increasing of synthesis time. Different morphologies were obtained by synthesis. The $Na_2Nb_2O_6.H_2O$ structures present fiber-like morphology and $NaNbO_3$ present cubic-like morphology, but after thermal treatment, the $Na_2Nb_2O_6.H_2O$ structures were converted to $NaNbO_3$ structures without changing the fiber-like morphology. The hysteresis loop indicated that improve of remnant polarization depends on the anisotropy particles.

ACKNOWLEDGMENTS

The authors thanks to the LMA-IQ for providing the FEG-SEM facilities, to Profº Drº Maximo Siu Li of São Carlos Physical Institute for providing the PL measurements Brazilian research funding agencies CNPq and FAPESP-CEPID/CDMF 2013/07296-2 for providing the financial support of this research project.

REFERENCES

1. C.A. Oliveira, E. Longo, J.A. Varela, M.A. Zaghete, *Ceram. Int.* **40** 1717 (2014)
2. C. Lizandara-Pueyo, S. Siroky, M. R. Wagner, A. Hoffmann, J. S. Reparaz, M. Lehmann, S. Polarz, *Adv. Funct. Mater.* **21** 295 (2011)
3. G.F. Teixeira, G. Gasparotto, E.C. Paris, M.A. Zaghete, E. Longo, J.A. Varela, *J. Lumin.* **132** 46 (2012)
4. M. Boukriba, F. Sediri, N. Gharbi, *Mater. Res. Bull.* **48** 574 (2013)
5. T.-Y. Ke, H.-A. Chen, H.-S. Sheu, J.-W. Yeh, H.-N. Li, C.-Y. Lee, H.-T. Chiu, *J. Phys. Chem. C* **112** 8827 (2008)
6. H. Song, W. Ma, *Ceram. Int.* **37** 877 (2011)
7. A.J. Paula, M.A. Zaghete, E. Longo, J.A. Varela, *Eur. J. Inorg. Chem.* **8** 1300 (2008)
8. Y. Shiratori, A. Magrez, W. Fischer, C. Pithan, R. Waser, *J. Phys. Chem. C* **111** 18493 (2007)
9. B. Zielinska, E. Borowiak-Palen, R. J. Kalenczuk, *J. Phys. Chem. Solids* **72** 117 (2011)
10. A. Yu, J. Q, L. Liu, H. Pan, X. Zhou, *Appl. Surf. Sci.* **258** 3490 (2012)
11. H. Zhu, Z. Zheng, X. Gao, Y. Huang, Z. Yan, J. Zou, H. Yin, Q. Zou, S. H. Kable, J. Zhao, Y. Xi, W. N. Martens, R. L. Frost, *J. Am. Chem. Soc.* **128** 2373 (2006)
12. A.J. Paula, R. Parra, M.A. Zaghete, *J.A. Varela, Mater. Lett.* **62** 2581 (2008)

Mater. Res. Soc. Symp. Proc. Vol. 1675 © 2014 Materials Research Society
DOI: 10.1557/opl.2014.878

High Transmission and Low Resistivity Cadmium Tin Oxide Thin Films Deposited by Sol-Gel

Carolina. J. Diliegros Godines[1], Rebeca Castanedo Pérez[1], Gerardo Torres Delgado[1] and Orlando Zelaya Ángel[2].

[1]Centro de Investigación y de Estudios Avanzados del Instituto Politécnico Nacional, Unidad Querétaro, A. P. 1-798, Querétaro, Qro. 76001, México.

[2]Depto. de Física, Centro de Investigación y de Estudios Avanzados del Instituto Politécnico Nacional, A.P. 14-740, México 07360, México D.F., México.

ABSTRACT

Transparent conducting cadmium tin oxide (CTO) thin films were obtained from a mixture of CdO and SnO_2 precursor solutions by the dip-coating sol-gel technique. The thin films studied in this work were made with 7 coats (~200 nm) on corning glass and quartz substrates. Each coating was deposited at a withdrawal speed of 2 cm/min, dried at 100°C for 1 hour and then sintered at 550°C for 1 hour in air. In order to decrease the resistivity values of the films, these were annealed in a vacuum atmosphere and another set of films were annealed in an Ar/CdS atmosphere. The annealing temperatures (Ta) were 450°C, 500°C and 550°C, as well as 600°C and 650°C, when corning glass and quartz substrates were used, respectively. X-Ray diffraction (XRD) patterns of the films annealed in a vacuum showed that there is only the presence of CTO crystals for 450°C≤ Ta ≤ 600°C and CTO+SnO_2 crystals for Ta=650°C. The films annealed in Ar/CdS atmosphere were only constituted of CTO crystals independent of the Ta. The minimum resistivity value obtained was ~4 x 10^{-4} Ωcm (R_{sheet}= 20 $\Omega/\square$) for the films deposited on quartz and annealed at Ta=600°C under an Ar/CdS atmosphere. The films deposited on quartz showed the higher optical transmission (~90%) with respect to the films deposited on corning glass substrates (~85%) in the Uv-vis region. For their optical and electrical characteristics, these films are good candidates as transparent electrodes in solar cells.

INTRODUCTION

Transparent conductive oxide films (TCO), such as Zn_2SnO_4, Cd_2SnO_4, $CdIn_2O_4$, SnO_2 and ZnO have found wide applications as transparent electrodes in photovoltaic, flat-panel displays, electrochromic devices and active electrodes in photo-electrochemical solar cells [1-4]. However, cadmium tin oxide thin films have received the greatest attention due to their use as a TCO in a CdTe polycrystalline based thin-film solar cell with a total-area efficiency of 16.5% [2]. Cd_2SnO_4 has been grown by many techniques and thermally treated after deposition in atmospheres such as air [5], He [6], N [7], Ar [5], vacuum [8] at temperatures up to 700°C, showing an improvement in the electrical properties. However, thermal treatments in Ar/CdS atmosphere have shown the best annealing process to improve the electrical properties of cadmium tin oxide films. Meng et al. [9] reported resistivity values of 1.6 x 10^{-4} Ωcm for CTO films deposited by radio frequency magnetron sputtering and annealing treatment in Ar/CdS atmosphere at 600°C. Although, high quality CTO films has been obtained by sputtering [5, 6, 8], a post deposition treatment or high temperatures of deposition are necessary to obtain a single crystalline phase of Cd_2SnO_4. In this paper, cadmium tin oxide films were deposited by the sol-gel dip-coating technique using a sintered temperature (Ts) of 550°C. After, thermal treatments were carried out at different annealing temperatures under two different atmospheres, Ar/CdS and vacuum. In this work, the study of the behavior of the films was made at different temperatures in order to achieve the best electrical, optical and structural properties.

EXPERIMENTAL DETAILS

Cd_2SnO_4 films

The cadmium stannate precursor solution was made from the mixture of the cadmium oxide and tin oxide solutions obtained separately at room temperature (RT). In order to obtain, pure Cd_2SnO_4 both solutions were mixed at Cd/Sn ratio of 2.44. The cadmium oxide precursor solution was prepared using cadmium acetate (($CH_3COO)_2Cd\cdot2H_2O$) (1 mol), methanol (33 mol), glycerol (0.2 mol) and triethylamine (0.5 mol) [10]. The procedure of the tin oxide precursor solution was similar to the one previously reported [11], only the molar concentration of triethylamine was changed. The SnO_2 precursor solution was prepared starting from stannous chloride ($SnCl_2\cdot2H_2O$) (1 mol), ethanol (40 mol), glycerol (0.20 mol) and triethylamine (0.10 mol). Lactic acid (0.4 mol) was added to the mixture of both solutions in order to obtain a transparent final precursor solution. The films were deposited by the multiple-dipping method on commercial glass substrates (Corning 2947) and quartz substrates, 24 h after the preparation of the precursor solution. The withdrawal speed was 2.0 cm/min. All the films were first thermally pretreated at 100°C and then subjected to a sintering treatment at Ts=550°C, in both cases in an air atmosphere for 1 h. The films were constituted of seven coats. Finally, a set of films were annealed in vacuum for 1h changing Ta from 450°C to 650°C in steps of 50°C. Another set of films was annealed at the same temperatures for 30 min in a quartz tube oven (2.5" diameter), where the ambient was Ar with CdS vapor at atmospheric pressure (Ar/CdS). CdS was introduced by positioning a 300 nm thick CdS-coated substrate to the surface of the CTO substrate with CdS film in direct contact with CTO. The CdS film was grown by chemical bath deposition, on glass and quartz substrates, using cadmium acetate (($CH_3COO)_2Cd\cdot2H_2O$, 0.031 mol), ammonium acetate ($C_2H_7NO_2$, 1.081 mol), ammonium hydroxide (NH_3, 30 ml, 28-30%), Thiourea (CH_4N_2S, 0.246 mol) and water (550ml). For each treatment a new CdS film was used to guarantee the same CdS portion applied to the CTO films. This method of annealing was previously reported by T. Meng et al. [9]. It is important to mention that for the annealing treatments at Ta= 600°C and 650°C the substrate used was quartz, while for the treatments at Ta= 450°C, 500°C and 550°C the substrate was commercial Corning glass (2947).

Characterization

The thickness of the films was measured by means of an optical profilometer (Bruker ContourGT), after removing a step-like portion of them with HCl. The X-ray diffraction (XRD) patterns were recorded using a Rigaku Ultima IV diffractometer (CuKα1 radiation, 1.5406 Å), employing a thin film attachment. Measurements of atomic concentration of elements were achieved by means of the Electron Dispersion Spectroscopy (EDS) technique using a Phillips XL30 (ESEM). The ultraviolet–visible (UV–Vis) spectra of the films were measured on a Cary 5000 UV-Vis-NIR, in the 200 nm–1100 nm wavelength range, the measurements were made without the glass substrate as reference. The resistivity values were measured by Hall Effect in an Ecopia HMS-3000 using a magnetic field of 0.5 T. Indium contacts were used in the electrical measurements. All measurements were carried out at room temperature. Atomic Force Microscopy (AFM) imaging was performed at 2% of relative humidity and 15°C with a commercial SPM system, a Bruker/Veeco/Digital Instruments Nanoscope IV Dimension 3100, operating in tapping mode and using Budget Sensors Cr/Pt-coated silicon probes (ElectriTap300-G).

DISCUSSION

The Cd_2SnO_4 thin films are constituted of 7 coats with an average thickness of 190±10 nm. The thickness of the films does not show change as the temperature increases under the two different annealing processes. XRD patterns of the films annealed in vacuum and under an Ar/CdS atmosphere are shown in figure 1a and 1b, respectively. In the bottom of the Figure 1, the XRD pattern of the films only sintered at Ts=550°C is shown; where it can be observed that the crystalline phase of Cd_2SnO_4 is already achieved. When the films are annealed in vacuum (Figure 1a) from 450°C to 600°C, the films are only constituted by Cd_2SnO_4 crystals (PDF#34-0928). At Ta=650°C in vacuum, the films show the presence of Cd_2SnO_4 and SnO_2 (PDF#46-1088) crystals. It was previously reported that for high temperatures under different annealing atmospheres (He, Ar/H$_2$, N) a secondary phase of SnO_2 or Cd_2SnO_3 appeared [6, 5, 7]. The presence of a second phase at high temperatures can be associated with the Cd which is a very volatile element [9], this provokes a higher loss of Cd than Sn, leading to the formation of a second phase based on Sn-O bonds. On the other hand, the films annealed under an Ar/CdS atmosphere (Figure 1b) only show the presence of Cd_2SnO_4 crystals for all the annealing temperatures. An atmosphere rich in Cd prevents the volatilization of Cd, leading to a stable structure.

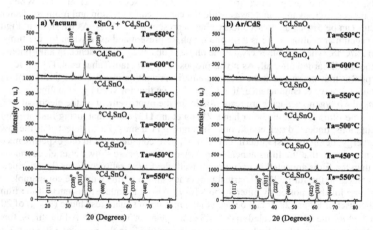

Figure 1. XRD patterns of cadmium tin oxide films annealed in a) vacuum and b) Ar/CdS atmosphere.

From the X-ray diffraction data the mean crystallite size (β) with (311) preferred orientation was calculated using Debye–Scherrer's formula [12], $\beta = 0.9\lambda$ / (FWHMcosθ), where λ is the wavelength of the X-rays used, FWHM is the full width half maximum of the corresponding peak and θ is the Bragg angle. The average crystallite size decreases as the temperature increases, from 39.0 nm to 32.0 nm when the films were annealed in Ar/CdS. While for the films annealed in vacuum, the crystallite size decreases from 41 nm to 30 nm as the temperature increases until 600°C, but for the films annealed at 650°C the crystallite size increases to 38 nm due to the multiphase condition of the film. The Cd/Sn ratio shows that for the films annealed in vacuum, there is a strong loss of Cd as the temperature increases from 1.8 to 1.4. This fact is more evident for the films annealed at 650°C (decreases from 1.4 to 0.9), which is in concordance with the XRD patterns where the films show the presence of $Cd_2SnO_4 + SnO_2$ crystals. This fact can be associated to a lower dissociation energy of the Cd-O bond (236 KJ/mol) than the ones of Sn-O bond (528 KJ/mol) by the presence of an annealing atmosphere and high temperatures [14,15]. For the films annealed in Ar/CdS, the ratio shows that the Cd rich environment does not let an important diminishing of the Cd/Sn ratio (~1.9). It is shown that the films annealed at Ta=600°C and 650°C the Cd/Sn ratio increases near to the stoichiometric ratio (2.15), while for the lower temperature it decreases from 1.9 to 1.8.

The electrical properties of the films only sintered at Ts=550°C in air do not show the adequate electrical properties for their use as a TCO ($\rho = 2.1 \times 10^{-2}$ Ωcm, $\mu = 6.32$ cm^2/Vseg and n= 4.65×10^{19} cm^{-3}). However, as it can be seen in Figure 2a, the minimum resistivity value of ρ~4 $\times 10^{-4}$ Ωcm was reached for the films deposited on quartz substrate and annealed at Ta=600°C in Ar/CdS atmosphere. The resistivity values decrease as the Ta increases up to 600°C, for both annealing treatments. On the other hand, both the carrier concentration and mobility increase, reaching the maximum values of 7.5 $\times 10^{20}$ cm^{-3} and 19 cm^2/Vseg, respectively, for the films annealed in Ar/CdS and Ta=600°C (see Figure 2b and 2c). When Ta=650°C, the resistivity values increases while the carrier concentration and mobility decrease. This behavior can be associated to secondary phases present in the films. As we show in Figure 1a, for the vacuum annealing, there is a secondary phase present and even when for the films annealed in Ar/CdS, where no other crystalline phase is observed, its presence in amorphous phases cannot be discarded after all. As it was proposed by Bel Hadj Tahar et al. [7], these secondary phases may segregate at the grain boundaries, resulting in a decrease in the mobility and consequently, a weak conductance. It is widely known that for vacuum annealing experiments, the oxygen vacancies are responsable for the improvement on the electrical properties of the films [7]. On the other hand, Zhang et al. [13] show that, using first principles defect calculations, under Cd rich conditions, Sn atoms on Cd sites (Sn$_{Cd}$) defects has the lowest formation energy, being the best kind of defect to improve conductivity. In our experiments of annealing, it is shown that the films annealed in Ar/CdS atmosphere show better electrical properties than the films annealed in vacuum. As well, the Cd/Sn ratio confirms the presence of higher Cd content in the films. The electrical properties shown in this work are comparable to the best values, previously reported by other authors [6-9]. Also, the square resistance of the films is shown in the inset of Figure 2a. It can be seen that a minimum square resistance (R$_{sheet}$) of 20 Ω/$\square$, is reached for the films annealed in Ar/CdS atmosphere at Ta=600°C. All the films, for both annealing treatments, show high transmittance around 85% in the range of 450 nm $\leq \lambda \leq$ 1100 nm. However, the films annealed in Ar/CdS atmosphere at 600°C and 650°C increase their average transmittance to 90%. In Figure 3, it is shown the transmission curve for the films annealed in Ar/CdS at Ta=600°C, which is the film with the best electrical properties (R$_{sheet}$=20

Ω/□). AFM topography images for the films annealed in Ar/CdS are also shown, where it can be seen that the aggregate size of the films is ~50 nm for the films at 600°C and ~100 nm for the films annealed at 650°C.

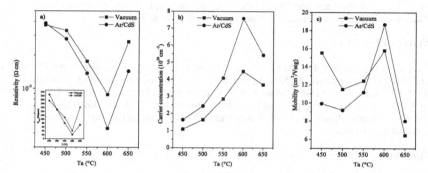

Figure 2. a) Resistivity, b) carrier concentration and c) mobility of the films as a function of the annealing temperature (Ta), for the films annealed in vacuum (■) and under an Ar/CdS atmosphere (●).

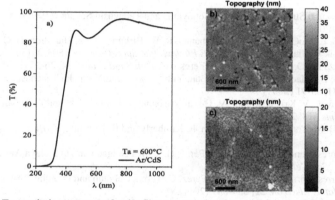

Figure 3. a) Transmission spectrum for the films annealed at Ta= 600°C in Ar/CdS atmosphere. AFM topography image of CTO film b) annealed at Ta= 600°C in Ar/CdS and c) annealed at Ta= 650°C in Ar/CdS.

CONCLUSIONS

In this work, we studied the behavior of cadmium stannate thin films under two annealing processes, vacuum and Ar/CdS atmospheres, at temperatures from 450°C to 650°C. The lowest resistivity value of 4×10^{-4} Ω cm and square resistance of 20 $\Omega/\square$ were obtained for the CTO films annealed in an Ar/CdS atmosphere at 600°C. All the films showed high transmittance around 85% in the 450 nm $\leq \lambda \leq 1100$ nm range. The films with thermal treatment in Ar/CdS at 600°C and 650°C show higher transmittance around 90% in the range of 450 nm$\leq \lambda \leq 1100$ nm. The optical and electrical characteristics of the films and the simplicity and economy of the technique used in their preparation, make them good candidates as TCO's.

AKNOWLEDGMENTS

This work was supported by the Secretaria de Ciencia, Tecnología e Innovación del Distrito Federal (SECITI) under project # 258/2012 and the Consejo Nacional de Ciencia y Tecnología (CONACYT) under Project FOMIX-Qro # 199228. The authors also thank CONACyT and CONCYTEQ for the fellowship awarded to M. Sci. Carolina Janani Diliegros Godines. The authors wish to thank M. Sci. Joaquín Márquez Marín, M. Sci. Cynthia I. Zúñiga Romero for their technical assistance.

REFERENCES

[1] C.S. Ferekides, R. Mamazza, U. Balasubramanian and D.L. Morel, *Thin Solid Films* **480–481**, 224-229 (2005).
[2] X. Wu, *Solar Energy* **77**, 803-814 (2004).
[3] M. A. Flores, R. Castanedo, G. Torres and O .Zelaya, *Sol. Energy Mater. Sol. Cells* **94**, 80-84 (2010).
[4] Prasad Manjusri Sirimanne, Noriyuki Sonoyama and Tadayoshi Sakata, *J. Solid State Chem.* **154**, 476-482 (2000).
[5]T. Meng, B. E. McCandless, W. A. Buchanan, R. W. Birkmire, C. T. Hamilton, B. G. Aitken and C. A. Kosik Williams, *Proc. 38th IEEE Photovoltaic Spec. Conf.*, 001803-001806 (2012).
[6] R. Mamazza Jr., D. L. Morel and C. S. Ferekides, *Thin Solid Films* **484**, 26-33 (2005).
[7] Radhouane Bel Hadj Tahar, Takayuki Ban, Yutaka Ohya and Yasutaka Takahashi, *J. Am. Ceram. Soc.* **84**, 85-91 (2001).
[8] R. Kumaravel, V. Krishnakumar, V. Gokulakrishnan, K. Ramamurthi and K. Jeganathan, *Thin Solid Films* **518**, 2271-2274 (2010).
[9] T. Meng, B. McCandless, W. Buchanan, E. Kimberly and R. Birkmire, *J. Alloy. Compd.* **556**, 39-44 (2013).
[10] M. A. Flores Mendoza, R. Castanedo Pérez, G. Torres Delgado and O .Zelaya Ángel, *Sol. Energy Mater. Sol. Cells* **93**, 28-32 (2009).
[11] D. Y. Torres Martínez, R. Castanedo Pérez, G. Torres Delgado and O. Zelaya Angel, *J. Mater. Sci: Mater. Electron.* **22**, 684-689 (2011).
[12] M. Birkholt, *Thin film analysis by X-Ray scattering*, first edition ,WILEY-VCH Verlag GmbH & Co. KGaA,(Weinheim, 2006) pp. 268-278.
[13]S. B. Zhang and S.-H. Wei, *Appl. Phys. Lett.* **80** 1376-1378 (2002).
[14] D. R. Lide, ed., CRC Handbook of Chemistry and Physics, 90th ed. (CD-ROM Version), CRC Press/Taylor and Francis, Boca Raton, FL, 2010.
[15] T. Meng, B. E. McCandless, W.A. Buchanan, R.W. Birkmire, C.T. Hamilton, B. G. Aitken, C. A. K. Williams, Proc. 38th IEEE Photovoltaic Spec. Conf., 2012, pp.001803-001806.

Mater. Res. Soc. Symp. Proc. Vol. 1675 © 2014 Materials Research Society
DOI: 10.1557/opl.2014.887

Facile One-Pot Synthesis of Rhenium Nanoparticles

Tuğçe Ayvalı[1], Pierre Lecante[2], Pier-Francesco Fazzini[3], Angélique Gillet[3], Karine Philippot[1] and Bruno Chaudret[3]

[1] CNRS ; LCC (Laboratoire de Chimie de Coordination) ; 205 Route de Narbonne ; F-31077 Toulouse (France) and UPS (Université de Toulouse), INPT, LCC ; F-31077 Toulouse (France)

[2] CNRS UPR 8011; CEMES (Centre d'Elaboration des Matériaux et d'Etudes Structurales) 29 Rue Jeanne Marvig, 31055 Toulouse (France)

[3] LPCNO (Laboratoire de Physique et Chimie des Nano-Objets) ; UMR 5215 INSA-CNRS-UPS, Institut des Sciences Appliquées, 135 Avenue de Rangueil, F-31077 Toulouse (France)

ABSTRACT

This paper describes the organometallic synthesis of pure rhenium nanoparticles (Re NPs) and their characterization by a combination of state-of-the art techniques (TEM, HAADF-STEM, EDX, WAXS, EA, FT-IR). The Re NPs synthesis is achieved by reducing the $[Re_2(C_3H_5)_4]$ complex in solution under a dihydrogen atmosphere and in the presence of hexadecylamine or polyvinylpyrrolidone as stabilizing agents. The so-obtained Re NPs are monodisperse with a mean size of 1.1 nm (0.3) nm and display a spherical shape with a disordered hcp structure.

INTRODUCTION

Nanomaterials of various types have experienced a significant development over the recent decades for their use in catalysis, to cite only one domain of application among many others [1]. However this development has concerned only a few classes of elements. For example, the preparation and characterization of nanoparticles of group 7 transition metals have been reported in very few articles [2].

Refractoriness, mechanical strength, high melting point and chemical resistance to be poisoned from N, S and P make rhenium as an attractive metal for aircraft engines and microelectronics. Rhenium has also been reported as having a positive contribution in terms of conversion and selectivity of complex catalytic processes such as glycerol reforming, hydrocarbon transformations and hydrogenation of difficult functional groups [3]. Surprisingly, there is a lack of information in the literature about the synthesis of pure rhenium nanoparticles (Re NPs). Although there are some examples [4-10], they all suffer from difficult synthetic protocols, polydispersity, high oxidation state and/or incomplete characterization. Thus, the preparation of well-controlled uniformly dispersed and reproducible Re NPs still remains challenge.

In our group, we have a wide experience in the synthesis of metal nanoparticles of various natures by using organometallic complexes as metal precursor [11]. This organometallic approach allows the formation of size-controlled metal nanoparticles with a clean surface (hydrides and chosen stabilizing agents are only present). Thus, we here report a facile and one-pot synthesis method to produce pure Re NPs by using [Re$_2$(C$_3$H$_5$)$_4$] complex as rhenium source. This method consists in the decomposition of [Re$_2$(C$_3$H$_5$)$_4$] in solution under a dihydrogen atmosphere, in the presence of hexadecylamine or polyvinylpyrrolidone as stabilizing agents. By this way, stable colloidal solutions containing monodispersed Re NPs with a narrow size distribution centered at 1.1 ± 0.3 nm were obtained. A combination of state-of-the art techniques (TEM, HAADF-STEM, EDX, WAXS, EA, FT-IR) as well as surface hydride quantification allowed us to characterize precisely the composition and surface state of these novel Re NPs. This paper is a part of a larger study presenting the complete characterization of these nanoparticles as well as the influence of reaction parameters on their morphology and surface reactivity studies [12].

EXPERIMENTAL DETAILS

General

All operations were carried out using Standard Schlenk tube and Fischer-Porter bottle techniques or in a glove-box under argon. All solvents were purified and degassed by freeze-pumping before using. Dirheniumtetraallyl [Re$_2$(C$_3$H$_5$)$_4$] was synthesized according to a slightly modified literature procedure [13].

Characterization

Re NPs were characterized by TEM and HAADF-STEM after deposition of a drop of colloidal solutions resulting from dissolution of purified samples into toluene or dichloromethane for Re/HDA and Re/PVP NPs respectively, over a covered holey copper grid inside the glove-box. TEM Analysis was performed at "*Service Commun de Microscopie Electronique de l'Université Paul Sabatier*" (TEMSCAN-UPS) by using JEOL JEM 1011 CX-T electron microscope operating at 100 kV with a point resolution of 4.5 Å. The approximation of the particle mean size was made through a manual analysis of enlarged micrographs by measuring at least 250 particles on a given grid. HAADF-STEM analysis was performed at TEMSCAN-UPS with JEOL JEM 2010 electron microscope working at 200 kV with a resolution point of 2.5 Å.

Wide-Angle X-Ray Scattering (WAXS) was performed at CEMES-CNRS. Purified samples of Re/HDA and Re/PVP NPs were sealed in 2.5 and 1.5 mm diameter Lindemann glass capillaries respectively. The samples were irradiated with graphite-monochromatized Mo K$_\alpha$ (0.071069) radiation and the X-ray intensity scattered measurements were performed using a dedicated two-axis diffractometer. Radial distribution functions (RDF) were obtained by Fourier Transformation of the reduced intensity functions.

Fourier Transform Infrared (FT-IR) spectra were recorded on a Perkin-Elmer GX2000 spectrometer in the range 4000-400 cm^{-1}. All samples were prepared as KBr pellets and measured under argon atmosphere inside a glove-box.

ICP analyses were performed at "*Institute des Sciences Analytiques, Department Service Central d'Analyse*" (CNRS) in Lyon.

Synthesis

In a typical experiment, the complex [Re$_2$(C$_3$H$_5$)$_4$] (2.85 mM) and polyvinylpyrrolidone (PVP, 90 wt. %) or hexadecylamine (HDA, 1 molar equiv.) were introduced into a Fischer-Porter reactor and dissolved in anisole. The resulting orange solution was pressurized to 3 bar of H$_2$ at room temperature (RT). Then the Fischer-Porter reactor was immersed into an oil-bath pre-heated to 120°C. The homogeneous solution was stirred for 2 days at this temperature leading to a stable dark brown colloidal solution. The colloidal solution was then cooled down to RT and the excess hydrogen was eliminated under vacuum. For PVP-stabilized Re NPs (Re/PVP NPs), the solvent was removed by vacuum and the dark brown solid was washed with pentane several times at RT. Then the fine powder was dried overnight under vacuum. For HDA-stabilized Re NPs (Re/HDA NPs), the volume of solvent was concentrated to 5 ml under vacuum and cold pentane was added. The nanoparticles were washed several times with cold pentane to provoke their precipitation and the supernatant was removed by cannula under Ar. The dark brown residue was dried overnight under vacuum at RT leading to a sticky sample in spite of various purification treatments. This is explained by the presence of long alkyl chain amine at the metal surface as previously observed with other metals like Ru, Pd and Pt [14] and used as it is. By Elemental Analysis (EA) 7.53 and 37.0 wt.% of Re were found for Re/PVP and Re/HDA NPs respectively which are close to expected theoretical values (10 and 40 wt% Re content for Re/PVP and Re/HDA NPs respectively).

Hydride Quantification

A solution of freshly prepared Re NPs in anisole as a solvent was submitted to 5 cycles of 1 min vacuum/ 1 min argon in order to eliminate any hydrogen dissolved. Then 5 molar equiv. of 2-norbornene was added and the reaction medium was heated to 80°C to overcome the high dissociation energy of Re-H bond. The conversion of 2-norbornene to norbornane was monitored by GC until having constant norbornane amount.

RESULTS AND DISCUSSION

Re NPs were easily synthesized in one-pot by reducing the organometallic complex $[Re_2(C_3H_5)_4]$ under dihydrogen atmosphere and in mild reaction conditions (3 bar H_2; 120°C). The stabilization of these Re NPs was achieved by adding PVP (steric stabilization) or HDA (electronic stabilization) giving rise to stable nanostructures. The nanoparticle structure and composition were identified by using complementary characterization techniques (TEM, HAADF-STEM, EDX, WAXS, FT-IR) in combination with surface hydride quantification.

Characterization by TEM, WAXS and FT-IR

Transmission electron microscopy (TEM) analysis of purified samples revealed the presence of non-agglomerated NPs of 1.2 (0.3) and 1.0 (0.3) nm mean sizes for Re/PVP and Re/HDA NPs, respectively (Figure 1). We also used the Scanning Transmission Electron Microscopy (STEM) mode with a High Annular Dark Field detector (HAADF) which is highly sensitive to the Z number. With this imaging mode, the carbon signal is lowered due to the great difference of the rhenium Z number over the Carbon one which allowed us to have more precise images of the Re NPs.

The crystalline structure of the Re NPs was investigated by wide-angle X-ray scattering (WAXS). The Radial Distribution Functions (RDFs) of Re NPs are shown in Figure 2. Coherence lengths, which are by principle average measurements of crystalline domains, of Re/PVP and Re/HDA NPs were measured as 1.2 nm and 1.0 nm respectively. These values are in good agreement with the mean sizes determined by TEM. In order to have deeper information on the atomic structure of the nanoparticles, the experimental RDFs of the Re NPs were compared with a theoretical model adopting hcp structure (spherical; 1.3 nm) and taking into account a single and large static disorder factor (Figure 2). All the distances from the experimental RDFs adequately match the ones derived from the 0.274 nm bond length in bulk rhenium. However the fast decay of the amplitude with distance indicates the presence of a short-range order indicative of low crystallinity.

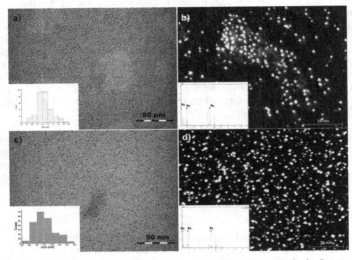

Figure 1. (a) and (c) TEM images of Re/HDA and Re/PVP NPs respectively. Insets show the size histograms of the corresponding images; (b) and (d) HAADF-STEM images of Re/HDA and Re/PVP NPs respectively. Insets correspond to EDX analysis of the nanoparticles.

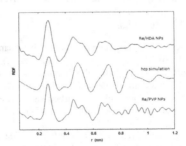

Figure 2. WAXS analysis of Re NPs (from top to bottom: Re/HDA NPs, hcp simulation for comparison, Re/PVP NPs.

FT-IR spectrum of Re/PVP NPs recorded on purified solid samples (Figure 3a) reveals the presence of absorption bands in the same region as for free PVP. Similarly, the FT-IR spectrum of Re/HDA NPs (Figure 3b) shows several peaks in the expected region for HDA which confirms the presence of amine ligand near the NP surface (bands at 3331 and 2849-2954 cm^{-1} : N-H and C-H stretching bands respectively; bands at 1584, 1256-1362, 1465 cm^{-1} : N-H,

C-N and C-H bendings). The decrease in N-H stretching band intensity and the shift in N-H bending band are attributed to σ-coordination of HDA on the NP surface.

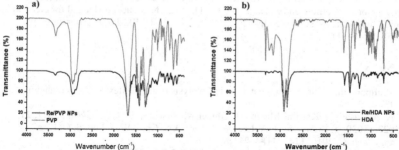

Figure 3. a) FT-IR Spectra of free PVP (red) and Re/ PVP NPs (black) ; b) FT-IR spectra of free HDA (red) and Re/HDA NPs (black)

Hydride Quantification

Since dihydrogen was used as reducing agent to synthesize the Re NPs, the presence of hydrides at their surface could be expected as observed with Ru NPs [15]. The amount of surface hydrides and their reactivity was monitored by a home-made titration method which is based on the hydrogenation of a simple olefin like norbornene without addition of dihydrogen. No significant hydrogenation of norbornene was observed at RT which can be explained by the strong interaction of hydrides with surface Re atoms given that Re-H bonds are very strong as in molecular chemistry [16]. To overcome high Re-H binding energy and activate the metallic surface, the experiment was performed at 80°C which led to the formation of norbornane. Assuming a mean size of 1.2 nm for Re/PVP NPs which thus have *ca.* 74% of surface Re atoms, we found *ca.* 1.1 hydride per surface Re atom. For Re/HDA NPs, we calculated 0.7 hydride per surface Re atom by assuming 1.0 nm size displaying *ca.* 81% surface atoms. These results show that hydrides are coordinated at the Re NPs surface and that heating is necessary to induce their reactivity.

CONCLUSIONS

To conclude, we describe in this paper an efficient method for the synthesis of novel, very small and monodispersed Re NPs (ca. 1-1.2 nm) stabilized either by a polymer (PVP) or a ligand (HDA) by using an organometallic complex as metal source and mild reaction conditions of solution chemistry. The resulting NPs display a spherical shape and adopt a highly disordered hcp structure. Surface hydrides are present on the surface and are very strongly coordinated to rhenium in agreement with the known molecular chemistry of rhenium. This work may open a new approach for the design of advanced Re-based systems such as bimetallic nanostructures as presently investigated in our group.

ACKNOWLEDGMENTS

The authors are grateful to Prof. K. Mertis for valuable discussions on the synthesis of the rhenium precursor. V. Collière and L. Datas at UPS-TEMSCAN are acknowledged for electron microscopy. This work was supported by CNRS, UPS and FP7-NMP2-Large program grant (Synflow 2010-246461).

REFERENCES

1. K. Philippot and P. Serp, Nanomaterials in Catalysis (Eds.), Wiley-VCH, Weinheim, 2013, **1**.
2. a) A. A. Ensafi, H. Karrimi-Maleh, M. Ghiaci, M. Arshadi, *J. Mater. Chem.,* 2011, **21**, 15022-15030; b) D. H. Lee, J. A. Lee, W. J. Lee, D. S. Choi, W. J. Lee, S. O. Kim, *J. Phys. Chem. C*, 2010, **114**, 21184–21189.
3. V. G. Kessler, G. A. Seisenbaeva, *Minerals*, 2012, **2**, 244-257.
4. M. R. Mucalo, C. R. Bullen, *J. Colloid Interface Sci.*, 2001, **239**, 71-77.
5. a) K. M. Babu, M. R. Mucalo, *J. Mater. Sci. Lett.*, 2003, **22**, 1755-1757; b) A. A. Revina, M. A. Kuznetsov, A. M. Chekmarev, *Dokl. Akad. Nauk*, 2013, **450**, 47-49.
6. a) N. Yang, G. E. Mickelson, N. Greenlay, S. D. Kelly, F. D. Vila, J. Kas, J. J. Rehr, S. R. Bare, *AIP Conf. Proc.*, 2007, **882**, 591; b) U. G. Hong, H. W. Park, J. Lee, S. Hwang, J. Yi, I. K. Song, *Appl. Catal. A*, 2012, **415**, 141-148; c) S. R. Bare, S. D. Kelly, F. D. Vila, E. Boldingh, E. Karapetrova, J. Kas, G. E. Mickelson, F. S. Modica, N. Yang, J. J. Rehr, *J. Phys. Chem. C*, 2011, **115**, 5740-5755.
7. a) C. Vollmer, E. Redel, K. Abu-Shandi, R. Thomann, H. Manyar, C. Hardacre, C. Janiak, *Chem.- Eur. J.*, 2010, **16**, 3849-3858; b) Y. Y. Chong, W. Y. Chow, W. Y. Fan, *J. Colloid Interface Sci.*, 2012, **369**, 164-169.
8. G. Y. Yurkov, A. V. Kozinkin, Y. A. Koksharov, A. S. Fionov, N. A. Taratanov, V. G. Vlasenko, I. V. Pirog, O. N. Shishilov, O. V. Popkov, *Composites: Part B*, 2012, **43**, 3192-3197.
9. C. D. Valenzuela, M. L. Valenzuela, S. Caceres, C. O'Dwyer, *J. Mater. Chem. A.*, 2013, **1**, 1566-1572.
10. a) J. Yi, J. T. Miller, D. Y. Zemlyanov, R. Zhang, P. J. Dietrich, F. H. Ribeiro, S. Suslov, M. M. Abu-Omar, *Angew. Chem. Int. Ed.*, 2013, **52**, 1-5; b) N. I. Buryak, O. G. Yanko, S. V. Volkov, *Ukr. Khim. Zh.*, 2013, **79**, 79-82.
11. C. Amiens, B. Chaudret, D. Ciuculescu-Pradines, V. Colliére, K. Fajerwerg, P. Fau, M. Kahn, A. Maisonnat, K. Soulantica, K. Philippot, *New J. Chem.,* 2013, **37**, 3374-3401.
12. T. Ayvalı, P. Lecante, P-F. Fazzini, A. Gillet, K. Philippot, B. Chaudret, *Chem Commun.*, 2014, **50**, 10809-10811.
13. A. F. Masters, K. Mertis, J. F. Gibson, G. Wilkinson, *New. J. Chem.*, 1977, **1**, 389-395.
14. a) C. Pan, K. Pelzer, K. Philippot, B. Chaudret, F. Dassenoy, P. Lecante, M-J. Casanove, *J. Am. Chem. Soc.*, 2001, **123**, 7584-7593; b) E. Ramirez, S. Jansat, K. Philippot, P. Lecante, M. Gomez, A. M. Masdeu-Bulto, B. Chaudret, *J. Organomet. Chem.*, 2004, **689**, 4601-4610; c) E. Ramirez, L. Eradés, K. Philippot, P. Lecante, B. Chaudret, *Adv. Func. Mater.*, 2007, **17**, 2219-2228.
15. F. Novio, K. Philippot, B. Chaudret, *Catal. Lett.*, 2010, **140**, 1-7.
16. D. Baudry, M. Ephritikhine, *J. Chem. Soc,. Chem. Comm.*, 1980, **6,** 249-250.

Mater. Res. Soc. Symp. Proc. Vol. 1675 © 2014 Materials Research Society
DOI: 10.1557/opl.2014.865

One-Step Microwave-Assisted Aqueous Synthesis of Silver-Based Nanoparticles Functionalized by Glutathione.

Myrna Reyes Blas[1], Maricely Ramírez- Hernandez[2], Danielle Rentas[1], Oscar Perales-Perez[1,3] and Felix R. Román[1].

[1] Department of Chemistry, University of Puerto Rico (UPRM), Mayaguez, USA
[2] Department of Chemical Engineering, University of Puerto Rico (UPRM), Mayaguez, USA,
[3] Department of Engineering Science & Materials, UPRM, Mayaguez, USA

ABSTRACT

The use of nano-sized silver and its alloys represents an interesting alternative to common food preservation methods, which are based on radiation, heat treatment and low temperature storage. These metal nanoparticles, embedded within a polymeric matrix for instance, would extend the shelf life of perishable foods while acting as a bactericidal agent to prevent food-borne illnesses. Common methods used in the synthesis of metal nanoparticles require toxic solvents and reagents that could be harmful to health and the food itself. In addition, several steps are required to obtain aqueous stable, i.e. dispersible, silver nanoparticles. In this work we propose the microwave-assisted aqueous synthesis of silver-based nanoparticles, (Ag Based NP) functionalized by glutathione (GSH) in a single-step using sodium sulfite (Na_2SO_3), as reducing agent. Ag-Based nanoparticles were synthesized at pH 6 and 1:3:1 ($AgNO_3$/GSH/ Na_2SO_3) molar ratio. UV-Vis measurement clearly showed the plasmon peak attributed to silver-based nanoparticles (374 nm). Highly monodispersed water stable Ag-based nanoparticles were observed and 3.897 ± 0.167 nm particle size was determined through Transmission Electron Microscopy. FT-IR measurements suggested the actual GSH-Ag based surface interaction through –SH and –COOH groups; the functionalization by GSH explained the high stability of the nanoparticles in aqueous suspensions. These Ag-GSH nanoparticles exhibited remarkable antimicrobial activity against E. Coli.

INTRODUCTION

One of the main causes of illness and death are the food-borne infectious diseases caused mainly by common bacterial species which have developed resistance to known antibiotics, E. coli is one of them[1]. Hence, it is imperative that there is development of new materials with antimicrobial properties which avoid contamination and propagation of these foods borne illnesses[2]. Although silver antimicrobial properties had been used for centuries, recent investigation recommended the use of silver and copper ions as superior disinfectants for wastewater generated from hospitals containing infectious microorganisms[3]. The synthesis procedures of silver nanoparticles are diverse spanning physical, chemical and biological methods[4-7]. These processes reported use of expensive and highly toxic reagents such sodium borohydride[8] and hydrazine[9]. There is literature that incorporates the use of several green reagents such as hemicellulose and starch. However, these methods follow several long and tedious steps, oftentimes taking from several hours to days [3,6,10,11]. In order to synthetize metal based nanoparticles intended to be used in biomedical, food and medical device packaging it is imperative that water stable nanoparticles can be produced. On the other hand, due to bacterial cell membrane pore size being in the order of nanometers, the nanoparticle size is an important synthesis parameter to consider and evaluate. Common methods used in the synthesis of metal nanoparticles require toxic solvents and reagents that could be harmful to health and the food itself. In addition, several steps are required to obtain aqueous stable, i.e. dispersible, silver nanoparticles. The use of microwave assisted synthesis of nanoparticles has been reported, high controlled temperatures can be reached and small size of nanoparticles can be obtained (3-20 nm) in a short time (seconds to minutes) [12]. In this work we propose the microwave-assisted aqueous synthesis of Ag Based nanoparticles, functionalized by glutathione (GSH) in a single-step using sodium sulfite (Na2SO3), as reducing agent.

EXPERIMENTAL

Materials
All reagents were of analytical grade and used without any further purification. $AgNO_3$ (ACS, ≥99%, Sigma-Aldrich), Na_2SO_3 (≥ 98%, Sigma-Aldrich), and L-Glutathione (≥ 98%, Sigma-Aldrich) were used for the synthesis of the AgBased nanoparticles.

Synthesis of AgBased Nanoparticles
The protocol using Glutathione (GSH) as reducing and capping agent is based on the work of *Baruwati et. al.* [12]. Here we had used $AgNO_3$ as a precursor salt of AgBasedNPs. The use of Na_2SO_3 as a reducing agent in addition to GSH was integrated based on the work of *Lopez-Miranda et. al.* [13]. A molar ratio of $AgNO_3$/GSH/ Na_2SO_3 for Ag Based NPs was prepared at pH 6. The synthesis was carried out during 10 min. at 115°C using a laboratory microwave furnace Mars6 Xtraction, CEM with power of 400W. Solids were coagulated with 2-propanol, centrifuged, dried in vacuum within an inert atmosphere at room temperature and characterized.

AgBased NP Characterization
AgBasedNP crystal structure was investigated using a Siemens D500 X-Ray Diffractometer with Cu K_α radiation. The morphology and size of the particles were examined by Transmission Electron Microscopy (TEM) and using a JEM-ARM200cF Transmission Electron Microscope. UV–VIS absorption spectra were recorded using a UV-Vis DU 800 spectrophotometer. The surface interactions of the nanoparticles were analyzed through a Perkin Elmer ATR Spectrum Two Fourier Transformed Infrared Spectrophotometer.

Bactericidal Capacity Assessment
The capacity for inhibition of bacterial growth was assessed in presence of AgBased NP with average crystallite sizes around 3.897 nm. AgBasedNP concentrations of 500, 1000, and 1500 µg/mL were evaluated. A fixed volume of 500 µL of the inoculum of *E.coli* at a concentration of 10^8 cells/mL was mixed with the different concentrations of nanocrystals, the final volume was filled up to 10 mL by using Tryptic Soy Broth (TSB) and incubated 24 hrs. at 37°C. A negative control was examined in absence of AgBasedNP. After that, a serial dilution from 10^{-1} to 10^{-8} was performed and 0.1 mL of each dilution was used to inoculate a petri dish and then incubate for 24 hrs. at 37°C. The colonies were counted in every petri dish and the Colony Forming Unit (CFU) per mL was established for every concentration of nanoparticles.

DISCUSSION

XRD Analyses

Although XRD analysis was performed, the diffraction angles could not be assigned in the XRD pattern for silver nanoparticles, this because less than 5nm particle size was observed. However, the results could be suggesting a mixture of silver nanoparticles and silver based nanoparticles (Data not shown). The AgBasedNP synthetized results were confirmed by UV spectrometry analysis later.

Transmission Electron Microscopy Analyses

Although Ag spherical-shaped NPs have been reported, the synthesis processes used require several steps and the particle sizes are larger than 10 nm[8]. Here, TEM and HRTEM images show spherical nanoparticles less than 5 nm (Figure 1a -1c). The nanometric nature of the AgBasedNP particles and their high crystallinity was evidenced (Figures 1 b-c) such as, high monodispersity and water solubility of nanoparticles. Figure 2 shows an AgBasedNP size distribution histogram where it can be observed that the largest particle population has a size between 3.0 and 4.5 nm. A mean of 3.897 ± 0.167 nm particle size was determined.

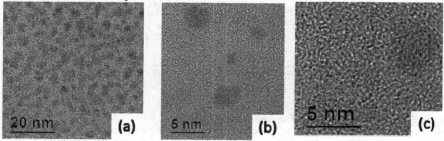

Figure 1. (a) TEM image of less than 5 nm AgBased nanocrystals; (b,c) HRTEM image showing the crystallographic planes of AgBased crystals.

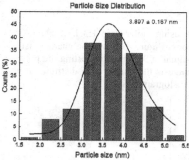

Figure 2. AgBasedNP Size distribution histogram. The particles are less than 5 nm mainly between 3.0 and 4.5 nm and with a mean size of 3.897 ± 0.167 nm.

Fourier Transformed - Infrared Spectroscopy Measurements

The interaction of Ag- based nanoparticles was evaluated using a Perkin Elmer ATR Spectrum Two FTIR spectrometer to verify the functionalization of particles. In the Ag-based nanoparticle spectra (Figure 3) the characteristic thiol peak (2520 cm^{-1}) of glutathione such as $-C=O$ (1704 cm^{-1}) was not observed, but the attachment of -COOH (1540, 1400 cm-1) can be observed in the spectra. These results suggest that the main interactions are through the metal nanoparticle and thiol and carboxylic groups from glutathione.

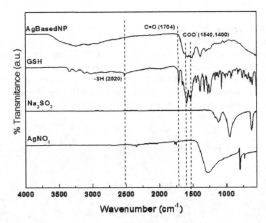

Figure 3. FT-IR spectra of GSH and Ag-based nanoparticles.
This figure shows the FT-IR spectra of glutathione, silver nitrate (AgNO$_3$), sodium sulfite (Na$_2$SO$_3$), glutathione (GSH) and Ag-based nanoparticles (Ag Based NP) with molar ratio Ag/GSH/Na$_2$SO$_3$ 1:3:1 at pH6. The Ag-based NPs spectra shows that the $-S-H$ (2520 cm^{-1}) peak was not observed and a displacement in – COOH (1540, 1400 cm^{-1}), such as, the attachment of $-C=O$ (1704 cm^{-1}).

UV-Vis Analyses

The synthesis of Ag Based NP using molar ratio 1:3:1 (AgNO$_3$/GSH/ Na$_2$SO$_3$) at pH 6 was evaluated by UV-Vis spectroscopy. Figure 4 shows the characteristic silver plasmon peak 350 nm and 450 nm with maximal signal at 374 nm, suggesting the presence of silver nanoparticles. Also, the narrow peak shape suggests the narrow size distribution which was demonstrated by TEM before in size distribution histogram (Figure 2).

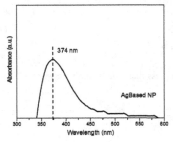

Figure 4. UV spectra of Ag Based NP. Characteristic silver plasmon peak can be observed (350 -450 nm).

Assessment of the Bactericidal Capacity

The bacterial growth inhibition of *E. coli* in presence of different concentrations of Ag Based NP was evaluated. Concentrations of 500, 1000, and 1500 µg/mL of Ag Based NP were considered. Evident bactericidal activity was observed in the cultures incubated in presence of Ag Based NP. In the presence of 500 µg/mL of Ag Based NP 62.7% of inhibition was observed, for 1000 µg/mL of Ag Based NP percentage of inhibition was increased to 90.7% and 100% of inhibition was observed when 1500 µg/mL of Ag Based NP was used. The result shows a remarkable bactericidal activity of Ag Based NP against *E. coli*. This bactericidal activity can be attributed to silver presence in nanoparticles.

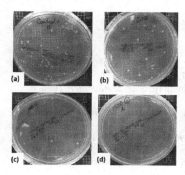

Figure 5. Culture in Plates. Each Petri dish shows the growth of *E. coli* colonies for 24 hrs at 37°C. (a) is the control culture incubated in absence of Ag Based NP, (b) culture in presence of 500 µg/mL of Ag Based NP, (c) 1000 µg/mL of Ag Based NP and (d) 1500 µg/mL of Ag Based NP. The 1500 µg/mL concentration 100% growth inhibition can be observed.

Table 1. Bactericidal Activity of AgBased Nanoparticles

Sample (µg/mL)	CFU**/mL of *E.coli*	Bacterial growth inhibition (%)
Control*	53.6×10^8	--
500	20×10^8	62.7
1000	5×10^8	90.7
1500	0	100

*Control was incubated in absence of AgBased NP.
** Colony-forming unit

CONCLUSIONS

Highly monodisperse and highly aqueous stable solution of AgBasedNPs were successfully synthesized using sodium sulfite as reductor and glutathione as a capping agent. A size of less than 5 nm of synthetized nanoparticles was determined. The formation of the spherical nanocrystal structure was demonstrated by TEM and HRTEM. ATR-FTIR spectroscopy techniques suggest that surface interactions between metal and capping agent (GSH) are through -SH and –COO⁻ groups

ACKNOWLEDGMENTS

This project is supported by the Agriculture and Food Research Initiative Competitive Grant No. 2012-67012-19806 from the USDA-National Institute of Food and Agriculture. The authors also thank the USDA-NIFA Center for Education and Training in Agriculture and Related Sciences (CETARS), Competitive Grant No. 2011-38422-30835. The TEM facility at FSU is funded and supported by the Florida State University Research Foundation, National High Magnetic Field Laboratory (NSF-DMR-0654118) and the State of Florida. The contribution from Professor Alicia Moradillos, Director of the Science and Technology Department, Antillean Adventist University at Mayaguez is also acknowledged.

REFERENCES

(1) CDC - A-Z Index for Foodborne Illness - Food Safety http://www.cdc.gov/foodsafety/diseases/ (accessed Apr 1, 2014).

(2) Rai, M.; Yadav, A.; Gade, A. *Biotechnol. Adv.* **2009**, *27*, 76–83.

(3) Ruparelia, J. P.; Chatterjee, A. K.; Duttagupta, S. P.; Mukherji, S. *Acta Biomater.* **2008**, *4*, 707–716.

(4) Hebbalalu, D.; Lalley, J.; Nadagouda, M. N.; Varma, R. S. *ACS Sustain. Chem. Eng.* **2013**, *1*, 703–712.

(5) Wang, Y.; Zheng, Y.; Huang, C. Z.; Xia, Y. *J. Am. Chem. Soc.* **2013**, *135*, 1941–1951.

(6) Prabhu, S.; Poulose, E. K. *Int. Nano Lett.* **2012**, *2*, 32.

(7) Majdalawieh, A.; Kanan, M. C.; El-Kadri, O.; Kanan, S. M. *J. Nanosci. Nanotechnol.* **2014**, *14*, 4757–4780.

(8) Joseph, S.; Mathew, B. *J. Mol. Liq.* **2014**, *197*, 346–352.

(9) Guzman, M.; Dille, J.; Godet, S. *Nanomedicine Nanotechnol. Biol. Med.* **2012**, *8*, 37–45.

(10) Peng, H.; Yang, A.; Xiong, J. *Carbohydr. Polym.* **2013**, *91*, 348–355.

(11) Taheri, S.; Baier, G.; Majewski, P.; Barton, M.; Förch, R.; Landfester, K.; Vasilev, K. *J. Mater. Chem. B* **2014**, *2*, 1838–1845.

(12) Baruwati, B.; Polshettiwar, V.; Varma, R. S. *Green Chem.* **2009**, *11*, 926–930.

(13) López-Miranda, A.; López-Valdivieso, A.; Viramontes-Gamboa, G. *J. Nanoparticle Res.* **2012**, *14*, 1–11.

Mater. Res. Soc. Symp. Proc. Vol. 1675 © 2014 Materials Research Society
DOI: 10.1557/opl.2014.879

Temperature Dependent Electrical and Dielectrics Properties of Metal-Insulator-Metal Capacitors with Alumina-Silicone Nanolaminate Films

Santosh K. Sahoo,[1] Rakhi P. Patel,[1] and Colin A. Wolden[1]

[1]Department of Chemical and Biological Engineering, Colorado School of Mines, Golden, Colorado 80401, USA

ABSTRACT

Alumina-silicone hybrid nanolaminate films were synthesized by plasma enhanced chemical vapor deposition (PECVD) process. PECVD allows digital control over nanolaminate construction, and may be performed at low temperature for compatibility with flexible substrates. These materials are being considered as dielectrics for application such as capacitors in thin film transistors and memory devices. Temperature dependent electrical and dielectric properties of the nanolaminate dielectric films in metal-insulator-metal structures are taken in the range of 200-340 K to better asses their potential applications for different devices. It is observed that the frequency dependent dielectric constant (ε_r) and ac conductivity (σ_{ac}) increase with the temperature. Both quadratic (α) and linear (β) voltage coefficient of capacitance (VCC) increases as the temperature increases. The temperature co-efficient of capacitance (TCC) decreases from 894 to 374 ppm/K as the Al_2O_3 composition increases in the alumina/silicone nanolaminates. Activation energy (E_a) for hopping conduction mechanism varies from 0.011 eV to 0.008 eV as the alumina composition increases from 50 to 83.3%.

INTRODUCTION

There are many studies about the dielectric and electrical properties of different thin films because of their importance for various electronic device applications. There is an increasing need to develop high quality metal-insulator-metal (MIM) capacitors for both analog and mixed-signal applications. Ideally these materials provide a high capacitance density as well as low leakage current density under the conditions of operation. Low dielectric loss (tanδ) is critical for RF and microwave device applications [1], while a low voltage co-efficient of capacitance (VCC) has been targeted by the international technology roadmap for semiconductors (ITRS) for analog applications [2,3]. Meeting all of the above requirements with a single material is proving to be a formidable challenge, particularly for applications requiring flexibility. The temperature and frequency dependent study of dielectric constant and ac conductivity (σ_{ac}) is very important to examine the charge carrier conduction mechanism in the insulators. Capacitors on flexible substrates often employ multilayer structures comprised of polymer and inorganic layers to provide flexibility and dielectric strength, respectively [4]. These hybrid organic-inorganic dielectrics are being pursued to serve a number of roles to enable flexible opto-electronics [5,6]. Thin film transistors (TFTs) are an important component in low cost, large area electronics such as flat panel displays and smart cards. The quality of the dielectric layer can significantly impact TFT performance, particularly when using organic semiconductors [7-9]. The two materials

(A,B) are typically arranged in either a sandwich structure (ABA) [7-9] or in the form of nanolaminates (AB-AB-AB-) [10,11].

There have been extensive investigations to better understand the mechanisms controlling dielectric performance in oxide based multi-layers (ZrO_2-Al_2O_3, TiO_2-Al_2O_3, HfO_2-Al_2O_3,TiO_2-Y_2O_3, $ZrLaO_x$-$ZrTiO_x$) [2, 12-16], but very limited studies on the performance of hybrid organic-inorganic dielectrics. We have recently demonstrated that silicone-alumina nanolaminates are promising candidates for hybrid dielectrics [11,17]. Both layers are deposited by plasma-enhanced chemical vapor deposition (PECVD) in a single chamber, which affords process simplification [18,19]. Silicone-alumina nanolaminates with ≥50% alumina content display excellent dielectric performance.

In this study, we focus on the frequency and temperature dependence of the dielectric properties such as dielectric constant, ac conductivity (σ_{ac}) and electrical properties as voltage co-efficient of capacitance (VCC), temperature co-efficient of capacitance (TCC), which is critical for various device applications. Detailed studies of the ac conductivity and hopping conduction mechanism of charge carriers has been reported for several inorganic-based multilayer dielectrics [1,20], but to date there have been no reports that have examined the hopping conduction process in hybrid multilayers comprised of alternating polymer and oxide thin films. With respect to voltage linearity, MIM capacitors are generally evaluated by characterizing the voltage coefficient of the capacitance (VCC) from the capacitance–voltage (C–V) characteristics. The C–V curves are approximately described as $C(V) = C_0(\alpha V^2 + \beta V + 1)$, [21] where C_0 is the capacitance at zero bias, and α and β are the quadratic and linear VCCs, respectively.

Here, we report on the dielectric and electrical properties of the high performance alumina/silicone nanolaminates. Frequency dependent dielectric constant (ε_r) and ac conductivity (σ_{ac}) at different temperatures are employed to estimate the activation energy (E_a) for hopping conduction mechanism in the nanolaminate films. Temperature dependent capacitance versus voltage (C-V) measurements are taken to assess their suitability as MIM capacitors for analog applications.

EXPERIMENTAL

Alumina-silicone hybrid nanolaminates were deposited on fluorine doped tin oxide (FTO, TEC-15) coated glass slides by PECVD process. The reactor used to deposit the nanolaminates in this work is a parallel plate, capacitively coupled PECVD system. The precursors for the individual layers were trimethyl aluminum (TMA, $Al(CH_3)_3$) and hexamethyl disiloxane (HMDSO). Alumina was deposited with the continuous delivery of TMA and O_2 by pulsed PECVD under self-limiting growth conditions at a rate of 3A/pulse whereas continuous wave PECVD was used to deposit silicone from HMDSO at a rate of 1.7 nm/s. The substrate temperature was kept at 105 °C for nanolaminate deposition. For the nanolaminates described in this work the dyad thickness was fixed at 30 nm, and the alumina volume fraction was varied from 50 – 83%. Each nanolaminate was an 11-layer structure comprised of 5 dyads along with an extra Al_2O_3 layer.[17] Thermal evaporation of aluminum contacts (D = 840 μm) was used to produce the MIM structures employed for electrical and dielectric measurements. The temperature dependent capacitance versus frequency (C-f), conductance versus frequency (G-f) in the frequency range of 1 kHz to 1 MHz, and capacitance versus voltage (C-V) in the voltage range of -15 to 0.154 V were measured in the temperature range of 200 to 340 K using an Agilent 4294A impedance analyzer and a cryostat.

RESULTS AND DISCUSSION

Figure 1 shows the variation of the dielectric constant with frequency in the various samples at different temperatures. The dielectric constant at 100 kHz is about 8.5 and 4.2 for the individual alumina and silicone films, respectively [22]. The experimental dielectric constant values obtained for the nanolaminates are in very good agreement with the theoretical calculated values by treating the individual layers as capacitors in series. The dielectric constant is more stable over the frequency range explored for the 50% Al_2O_3 nanolaminates compared to that of nanolaminates having 83.3% Al_2O_3. The more frequency stable dielectric properties observed for the 50% Al_2O_3 nanolaminates is due to the presence of more volume fraction of silicone polymers in comparison to 83.3% Al_2O_3 nanolaminates. Silicone has additional features that benefit these applications. Silicone polymers are largely nonpolar, as the methyl side groups prevent Si-O dipoles from approaching each other too closely. As a result, the intermolecular forces are weak and mainly composed of van der Waals interactions that decrease with the square of the distance between molecules. Consequently, electrical properties like dielectric constant are little affected over a wide range of frequency [23].

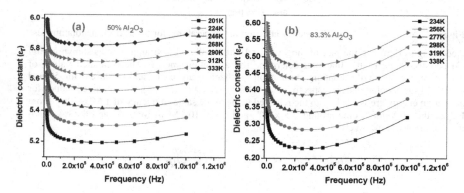

Figure 1. Dielectric constant vs. frequency at different temperatures for (a) 50% Al_2O_3 and (b) 83.3% Al_2O_3 nanolaminates.

From figure 1, it is also observed that the dielectric constant increases with the increase of temperature for both 50 and 83.3% Al_2O_3 nanolaminates. Generally, when the temperature increases, the structure relaxes and the orientation of the polarized units become easier and hence the value of orientational polarization increases. As a result, the total polarization of the system increases. Therefore, the dielectric constant increases with the rise of temperature as expected [20].

Figure 2 shows the frequency dependence of the ac conductivity (σ_{ac}) at different temperatures for alumina/silicone nanolaminates having 50% Al_2O_3 composition. It is observed that the ac conductivity increases as the temperature increases because of the flow of more charge carriers. The Arrhenius equation for the temperature dependence of the ac conductivity is given by [20],

$$\sigma_{ac} = \sigma_0 \, Exp\left(-E_a / kT\right) \tag{1}$$

where, σ_0 is the pre-exponential factor, E_a is the thermal activation energy for the hopping of charge carriers from one state to the next state, and T is the device temperature.

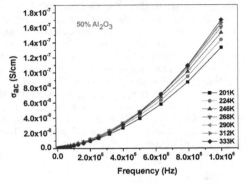

Figure 2. AC conductivity vs. frequency at different temperatures for 50% Al$_2$O$_3$ nanolaminates.

According to Equ^n (1), the plot of lnσ_{ac} versus 1000/T will give a straight line as shown in Fig. (3) and the slope of the line will determine the activation energy. The estimated values of activation energy for the hopping conduction mechanism are 0.011 and 0.008 eV for 50 and 83.3% Al$_2$O$_3$ nanolaminates, respectively at 1 MHz frequency (Fig. 3). The obtained values for activation energy are very close to the reported value at the same frequency of 1 MHz [20]. From Fig. 3, the σ_0 values come around 2.5 x 10^{-7} and 2.1 x 10^{-7} S/cm for 50% and 83.3% Al$_2$O$_3$ nanolaminates, respectively.

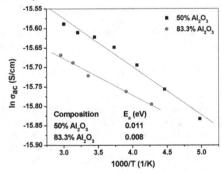

Figure 3. ln σ_{ac} vs. 1000/T for alumina-silicone nanolaminates with 50 and 83.3% Al$_2$O$_3$ composition.

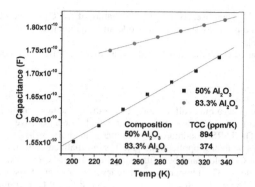

Figure 4. Capacitance vs. temperature at 100 kHz frequency.

Figure 4 shows capacitance versus temperature at 100 kHz frequency for the nanolaminates having 50 – 83.3% Al_2O_3 compositions. It is observed that the capacitance increases with the increase of temperatures for both nanolaminates. The temperature co-efficient of capacitance (TCC) is given by [1],

$$TCC = \frac{1}{C(T_1)} \frac{\Delta C}{\Delta T} \tag{2}$$

where, ΔC is the change in capacitance with respect to $C(T_1)$ and ΔT is the change in temperature relative to T_1, $C(T_1)$ is the capacitances at temperature T_1. The estimated TCC value varies from 894 to 374 ppm/K as the Al_2O_3 composition increases from 50 to 83.3%. The decrease in TCC values is due to the addition of more oxide layer compared to polymer layer. The obtained TCC values for the alumina-silicone nanolaminates in this study are very low and hence superior compared to the reported value of 2656 ppm/K for the single layer Al_2O_3 film with the similar thickness of 165 nm [20]. The low value of TCC observed for the nanolaminate films is due to the interface effect between two different materials in a multilayered structure. As the temperature is increased, the intermolecular forces between polymer chains in organic materials is broken which enhances the thermal agitation. Hence, at higher temperature the dielectric constant for polymer insulators is reduced due to increase in thermal motion which disturbs the orientation of the dipoles [24]. Therefore, the TCC value for silicone, a polymer insulator, is negative whereas this value is positive for Al_2O_3. As a result, the TCC value of the hybrid nanolaminates can be engineered by tuning the composition of the two individual materials in the multilayered structure. Therefore, nanolaminate films show temperature stable dielectric properties in comparison to single layer films. This temperature stability is very important, since in microwave devices the dielectric properties must be stable to ensure the antenna performance consistent and reliable with respect to temperature variation in the environments [1].

Lastly, we examine the voltage dependence of these capacitors to assess their potential in analog devices. Figure 5(a) shows the $\Delta C/C_0$ versus voltage (obtained from C-V plot) characteristics for alumina/silicone MIM capacitors. For the nanolaminates the voltage was swept from -15 V to 0.154 V. The observed asymmetric nature of $\Delta C/C_0$ -V characteristic is due to the different work functions for top/bottom electrode and asymmetric band structures of the nanolaminate dielectric [15].

At a particular frequency, the normalized capacitance of the MIM capacitor can be given by [25]

$$\frac{\Delta C}{C_0} = \frac{C(V) - C_0}{C_0} = \alpha V^2 + \beta V \tag{3}$$

where C_0 is the capacitance at zero bias voltage and C(V) is the capacitance at bias voltage, V, at a particular frequency, and α and β are the quadratic and linear voltage co-efficients of capacitance, respectively. The ITRS Roadmap suggests that a quadratic voltage co-efficient of α < 100 ppm/V^2 is required for analog applications [2]. The normalized capacitance versus bias voltage was extracted from C-V plots and best fits using Equation (3) are shown in Figure 5(a) for 50% Al_2O_3 nanolaminates. The temperature dependence C-V characteristics are taken for 50% Al_2O_3 nanolaminates. The estimated values of α and β at different temperatures are plotted in Figure 5 (b). It is observed that both α and β values increase with the increase of temperature. This phenomenon can be explained by the fact that, at higher temperature, higher electron concentration in the nanolaminate dielectric can be achieved because of a lower electron injection barrier at the electrode/dielectric interface, and it causes smaller relaxation time which consequently leads to a larger capacitance variation, according to a free-carrier injection model [26].

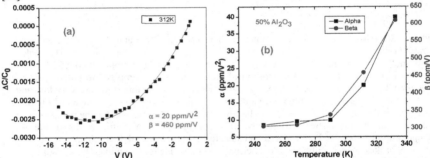

Figure 5. (a) $\Delta C/C_0$ vs V at 312K and (b) temperature dependence of quadratic VCC (α) and linear VCC (β) for the nanolaminates having 50% Al_2O_3 composition.

CONCLUSIONS

In conclusion, the frequency and temperature dependence of the dielectric properties for high performance alumina-silicone dielectrics have been quantified. The activation energy for hopping conduction mechanism in hybrid nanolaminates decreases as the composition varies from 50 to 83.3% Al_2O_3. Nanolaminate hybrid films show more temperature stable dielectric properties in comparison to that of single layered Al_2O_3 films. Combined with the high capacitance density (30 nF/cm^2), low dielectric loss, extremely low leakage current density (~10⁻

^{9} A/cm^2 at 1MV/cm), and having both temperature and frequency stable dielectric properties, these results demonstrate that alumina-silicone nanolaminates are promising candidates to serve in MIM capacitors on flexible substrates for analog applications in both high frequency and temperature variant environments.

ACKNOWLEDGEMENTS

We gratefully acknowledge the National Science Foundation for the support of this work through awards CMMI-0826323 and CBET 1033203. We would like to thank the National Renewable Energy Laboratory, Golden, Colorado for using their electrical and dielectric measurement facilities.

REFERENCES

1. S. K. Sahoo, D. Misra, M. Sahoo, C. A. MacDonald, H. Bakhru, D. C. Agrawal, Y. N. Mohapatra, S. B. Majumder, and R. S. Katiyar, J. Appl. Phys. **109**, 064108 (2011).
2. S. J. Kim, B. J. Cho, M. -F. Li, S.-J. Ding, C. Zhu, M. B. Yu, B. Narayanan, A. Chin, and D.-L. Kwong, IEEE Electron Dev. Lett. **25**, 538 (2004).
3. K. C. Chiang, C. -C. Huang, G. L. Chen, W. J. Chen, H. L. Kao, Y. -H. Wu, and A. Chin, IEEE Trans. Electron. Devices **53**, 2312 (2006).
4. R. P. Patel and C. A. Wolden, Appl. Surf. Sci. **268**, 416 (2013).
5. R. P. Ortiz, A. Facchetti, and T. J. Marks, Chem. Rev. **110**, 205 (2010).
6. M. C. Choi, Y. Kim, and C. S. Ha, Prog. Polym. Sci. **33**, 581 (2008).
7. A. L. Deman, M. Erouel, D. Lallemand, M. Phaner-Goutorbe, P. Lang, and J. Tardy, J. Non-Cryst. Solids **354**, 1598 (2008).
8. D. K. Hwang, C. S. Kim, J. M. Choi, K. Lee, J. H. Park, E. Kim, H. K. Baik, J. H. Kim, and S. Im, Adv. Mater. **18**, 2299 (2006).
9. Y. G. Seol, H. Y. Noh, S. S. Lee, J. H. Ahn, and N. -E. Lee, Appl. Phys. Lett. **93**, 013305 (2008).
10. L. D. Salmi, E. Puukilainen, M. Vehkamaki, M. Heikkila, and M. Ritala, Chem. Vapor Depos. **15**, 221 (2009).
11. S. K. Sahoo, R. P. Patel, and C. A. Wolden, Appl. Phys. Lett. **101**, 142903 (2012).
12. Y. H. Wu, C. K. Kao, B. Y. Chen, Y. S. Lin, M. Y. Li, and H. C. Wu, Appl. Phys. Lett. **93**, 033511 (2008).
13. J. C. Woo, Y. S. Chun, Y. H. Joo, and C. I. Kim, Appl. Phys. Lett. **100**, 081101 (2012).
14. S. K. Lee, K. S. Kim, S. W. Kim, D. J. Lee, S. J. Park, and S. Kim, IEEE Electron. Dev. Lett. **32**, 384 (2011).
15. Y. -H. Wu, C. -C. Lin, Y. -C. Hu, M. -L. Wu, J. -R. Wu, and L. -L. Chen, IEEE Electron. Dev. Lett. **32**, 1107 (2011).
16. L. L. Chen, Y. H. Wu, Y. B. Lin, C. C. Lin, and M. L. Wu, IEEE Electron Dev. Lett. **33**, 1447 (2012).
17. R. P. Patel, D. Chiavetta, and C. A. Wolden, J. Vac. Sci. Technol. A **29**, 061508 (2011).
18. R. P. Patel and C. A. Wolden, J. Vac. Sci. Technol. A **29**, 021012 (2011).

19. M. T. Seman, D. N. Richards, P. Rowlette, and C. A. Wolden, Chem. Vap. Deposition **14**, 296 (2008).

20. D. Deger, K. Ulutas, and S. Yakut, J. Ovon. Res. **8**, 179 (2012).

21. R. K. Hester, K. -S. Tan, M. de Wit, J. W. Fattaruso, S. Kiriaki, and J. R. Hellums, IEEE J. Solid-State Circuits **25**, 173 (1990).

22. S. K. Sahoo, R. P. Patel, and C. A. Wolden, J. Appl. Phys. **114**, 074101 (2013).

23. A. Von Hippel, *Dielectric Materials and Applications*. (John Wiley & Sons Inc., New York, 1954).

24. http://cdn.intechopen.com/pdfs-wm/39574.pdf

25. H. Hu, C. Zhu, Y. F. Lu, M. F. Li, B. J. Cho, and W. K. Choi, IEEE Electron. Dev. Lett. **23**, 514 (2002).

26. S. J. Ding, H. Hu, H. F. Lim, S. J. Kim, X. F. Yu, C. X. Zhu, M. F. Li, B. J. Cho, D. S. H. Chan, and S. C. Rustagi, IEEE Electron Dev. Lett., **24**, 730 (2003).

Mater. Res. Soc. Symp. Proc. Vol. 1675 © 2014 Materials Research Society
DOI: 10.1557/opl.2014.801

Soft Lithographic Printing of Titanium Dioxide and the Resulting Silica Contamination Layer

Travis Curtis, Lakshmi V. Munukutla and Arunachalanadar M. Kannan
Department of Engineering, Arizona State University, Mesa, Arizona, USA.

ABSTRACT

Soft lithographic printing techniques can be used to print nanoparticle dispersions with relative ease while allowing for a measureable degree of controllability of printed feature size. In this study, a Polydimethylsiloxane (PDMS) stamp was used to print multi-layered, porous, nanoparticle dispersions of titanium dioxide (TiO2), for use in a dye-sensitized solar cell application. The gelled patterns were then sintered and the surface of the printed sample was chemically analyzed.

X-ray photoelectron spectroscopy (XPS) was used to determine the surface constituents of the printed sample. The presence of a secondary peak feature located approximately 2.8 eV above the high resolution O1s core level binding energy peak was attributed to a contamination layer. Fourier transform infrared spectra (FTIR) of the printed sample revealed the presence of vibrational modes characteristic of the asymmetric bond stretching of silica, located at approximate wavenumbers of 1260 and 1030 cm^{-1}.

Soft lithographic techniques are a viable manufacturing technique in a number of disciplines and sintered nano-oxide dispersions are readily used as reaction centers in a number of technologies. The presence of a residual, bonded silicate contamination layer may preclude the soft lithographic printing of chemically active oxide surfaces.

INTRODUCTION

There are a multitude of fabrication processes that have been explored to control the dimensionality of materials on length scales of several micrometers down to the nanometer [1,2]. More traditional lithographic methods were established to process semiconductor material and typically implement a photon, electron or ion source that radiates a base material to physically and chemically alter its dimensionality, while masking layers serve to protect those areas that are intended to remain un-etched.

A less traditional approach, nano-imprint lithography (NIL) is a soft lithographic technique. NIL is a quasi-non-mechanical deformation process that is generally low cost and thus serves as an easily accessible tool to fabricate micro- and nanostructures. Typically, a polymeric, patterned relief mold, a stamp, is used to transfer, or print, the base material. PDMS is widely used to fabricate stamps as it has relatively good mechanical stiffness and low surface energy [3]. These two features are instrumental in the transfer process as a high modulus necessitates a stiff mold that will replicate accurately and low surface energy is a requisite to transfer— as the material being transferred will preferentially adhere to the substrate more readily than the stamp face. Numerous groups have used variations of NIL and other soft-lithographic techniques to form controlled nanostructures, while a number of groups have, more specifically, had success creating nanostructures from TiO$_2$ solutions [4-7]. Throughout, successful material transfer was typically predicated by solvent content and the dispersions film thickness, variables controlled by material concentration, coating spin speed and time.

EXPERIMENT

Mold Making and Substrate Preparation

A PDMS mold was fabricated using a 10:1 mass ratio of polymer base to curing agent (Sylgard 184, Dow Corning Corp.). After aggressive, yet controlled mixing, the polymer mixture was placed in a desiccator for 30 minutes. Occasionally the vacuum was broken so as to agitate the escaping air bubbles ascending towards the mixtures surface. The PDMS mixture was poured over a grating (Edmund Optics) that sat atop a Si-wafer; a square acrylic mold formed a filling template around the grating and aided in controlling the final dimensionality of the pour. The Si-wafer, grating template, acylic mold and PDMS mixture were placed into a 60°C oven for 6 hours and allowed to cool overnight. Using a laser cutter, the polymerized PDMS mold was lightly etched with a grid of stamp outlines. A straight razor was then used to segment the stamps, following the hatched lines etched by the laser. Each individual stamp was then carefully removed. A number of materials were used as substrates in the stamping process, to include: cleaved silicon, borosilicate glass, and gold-coated slide glass. The substrate was cleaned by sonication in a bath of acetone for 10 minutes, followed by isopropanol for an additional 10 minutes.

Printing TiO2 Surfaces

The stamp was placed feature side up on a level surface. Approximately 12.5 µL of a purchased TiO_2 solution (Solaronix Ti-Nanoxide T/SC) was placed directly on the stamp face. A glass rod was used to disperse the solution onto the surface and the stamp was placed feature side down on the substrate. When conditions warranted, the stamp was lightly pressed so as to displace trapped air. The solution was allowed to dry for a minimum of 1 hour. The stamps were carefully peeled away (removed) to reveal the transferred, gelled pattern. The process was repeated in a similar fashion to create multiple layers. The substrates were then sintered in a 450°C furnace for 1 hr and allowed to cool over night. The stamping process is illustrated in figure 1; also depicted, an SEM image of a typical printed oxide sample cleaved and viewed from the side.

"Standard" TiO_2 samples were also prepared to establish a baseline to which the printed sample could be compared. Using the identical TiO_2 solution used to prepare the printed samples, the standard samples were prepared via a doctor-blading technique. A scotch-tape layer was laser etched with a cutout in the same dimensions as the projected, printed stamp area. The tape was adhered to the substrate and approximately 12.5 µL of the TiO_2 solution was placed in front of the tape cutout. A flexible plastic blade was used as a squeegee to spread the solution across the substrate surface. The solution was allowed to air dry for approximately 1 hour. The process was then repeated to create multiple layers. The substrates were then heated in a 450°C furnace for 1 hr and allowed to cool.

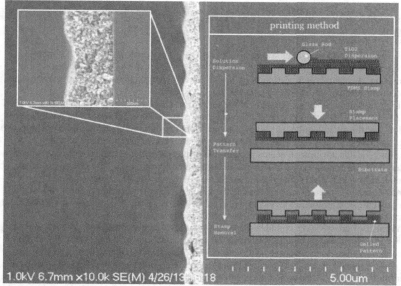

Figure 1. 1kV, 10k SEM image of deposited TiO2 layer viewed side-on (background image). Upper left: 1kV, 80k SEM image of deposited TiO2 layer. Inset to the right: the printing method used to form the printed oxide. This process was repeated to form multiple layers. Upon completion of the printing process the sample was sintered at 450C.

Sample Analysis

XPS was performed with a VG 220i-XL. The surface of the sample was charge compensated (1.51μA) to alleviate charge build-up. The acceleration voltage was 12kV (65W) and the spot size was 500 μm. The sample was placed perpendicular to the detector. The source was monochromatic Al-Kα (1486 eV), with a line width resolution of 0.8 eV. Spectra were deconvoluted with CasaXPS software.

A Hitachi S4700 Field Emission Scanning Electron Microscope (FESEM) was used to capture images of the TiO$_2$ samples. Imaging was typically conducted at low voltage (1kV) and a 6 mm working distance. Samples were evaluated from side-on and overhead. Side-on samples were typically cleaved through the center of the projected oxide area and placed in a sample holder for side-on viewing. Overhead samples were unaltered. Samples were uncoated and substrates were typically conducting.

Fourier Transform Infrared Spectroscopy (FTIR) was conducted on a standard sample and printed sample, each sintered on the same gold-coated glass substrate, to determine the presence of a contamination layer. The spectrum was taken at atmosphere and recorded over the range of 400-4000 cm^{-1}. The background spectrum was created using 64 signal averaged scans while the sample spectra were created using 24 signal averaged scans.

DISCUSSION

XPS Study

A survey spectrum was taken of the printed sample. The spectrum showed dominant O 1s and Ti 2p peaks, as well as pronounced C 1s and Si 2s and 2p peaks. Despite charge compensation, the sample was also initially shifted due to charge build-up. The spectrum was reoriented with respect to the adventitious carbon C 1s peak.

A high-resolution scan was taken near the O 1s region. Two peaks are present, one located at 529.26 eV and one located at 532.09 eV. The peak separation is 2.83 eV. This doublet formation, see figure 2, is quite perplexing, as one would not quite expect the radial symmetry of the O 1s shell to show any splitting. While there are a number of potentially plausible explanations for this attribute, it seems perhaps most likely that the abundance of oxygen and the secondary peak formation is due to a contamination layer brought on by the printing process. The stamp material used, PDMS, is a polymer consisting of a siloxane backbone with methyl groupings attached. It seems quite plausible then, that during the sintering process, the methyl groupings of the contamination layer that are attached to the gelled TiO_2 precursor surface would likely carbonize, leaving a Si-O linkage that would readily oxidize and form SiO_2.

A number of groups have conducted what amount to co-evolution studies of TiO_2 and SiO_2 films [8-13]. These studies tend to formulate the energy locations of the O1s peak(s), and the relative peak spacing, with respect to TiO_2 and SiO_2 concentration. While the methods are vast— photo-induced chemical vapor desorption, co-evaporation studies with simultaneous ion bombardment, coreshell structure studies, self-assembled monolayer studies, or plasma enhanced oxidation and evaporation studies— all the works capture snapshots of the O1s binding energy (BE) peak location with respect to SiO_2-TiO_2 concentration. Thus, they help establish bookend-like values for the O1s peak position, as they pertain to Si, Ti, and O concentration. Put another way, they seem to indicate samples with higher concentrations of SiO_2 show O1s peaks near approximately 533 eV and those samples that show higher concentrations of TiO_2 have accordingly an O1s peak that is at approximately 530 eV. Perhaps more importantly, as peak position is somewhat arbitrary, the referenced works help establish the difference in O1s BE. That is to say, the difference in energy between the oxygen that belongs to SiO_2 and the oxygen that belongs to TiO_2 and/or some hybrid-like TiO_2-silicate film, is established. The difference in O1s peak positions for the works referenced was between 1.8-2.9 eV. The results of this work establish a delta in the O1S BE, attributed to the displaced electrons originating in the oxygen belonging to SiO_2 and to TiO_2, at 2.8 eV.

A high-resolution scan was also taken near the Ti 2p region of the printed sample. Two peaks are present, one located at 455.12 eV and one located at 460.84 eV. The peak separation is 5.7 eV. The Ti 2p spectrum of the printed sample shows intrinsic Ti 2p spin-orbit coupling.

The ratio of degeneracy should be 2, as the total angular momentum, J, is related to the spin, s (+ or – ½). J= 1+s, so J = 1+ 1/2 or J= 1 – 1/2. J is then, J = 1/2 and 3/2. The peak intensity is proportional to the degeneracy, D= 2J+1. So, J= 3/2 yields a degeneracy of 4, while J= 1/2 yields a degeneracy 2. The expectation would be to see the intensity differ by a factor of 2 and upon inspection the calculated ratio of corrected intensities is 2.18 (4725/2172 = 2.18). While the ratio is near 2, the anti-parallel direction appears more favored. Also of note, there seems to be no underlying oxide or hydroxyl groupings, which would seem to suggest the printed surface is relatively, defect free.

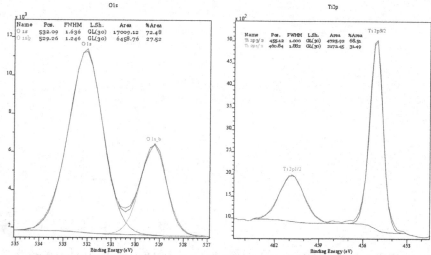

Figure 2. XPS high-resolution spectra of O1s and Ti2p region of printed sample, included also are curve-fits. *Left:* O1s peak, left-most peak is attributed to SiO_2 while secondary peak to the right is O1s attributed to TiO_2. Peak separation is approximately 2.8 eV. *Right:* Ti 2p peak with split-states (due to spin-orbit coupling), left-most peak is Ti 2p 1/2 while right most peak is Ti 2p 3/2. Ratio of degeneracy was calculated as 2.18, the anti-parallel spin direction appears more favored.

FTIR Study

FTIR spectra were taken of a standard sample (un-printed/no PDMS contact) and a printed sample. The juxtaposition of standard and printed sample spectra shows some significant differences in spectral features. A number of peaks, at specific wavenumbers, correlate to important markers that are intrinsic to SiO_2 bond states, and thus seem to point strongly towards the formation of a SiO_2 layer when printing, as can be seen in figure 3.

There is a peak structure present for both the printed and standard sample, at ~1630 cm^{-1}. This peak can be attributed to the bending mode of the hydroxyl group that is indicative of water present on the sample surface or within the localized atmosphere of the spectrograph [14]. The peak present in the spectra of the printed sample, at ~1260 cm^{-1}, can likely be attributed to the longitudinal optic (LO) mode of Si-O-Si asymmetric bond stretching, while the peak at ~1030 cm^{-1} is likely associated with the transverse optic (TO) mode of Si-O-Si asymmetric bond stretching [15-17]. There seems to also exist an inflection point at ~800 cm^{-1}. A peak at this wavenumber would seem to indicate the symmetric bond stretching of Si-O-Si [15].

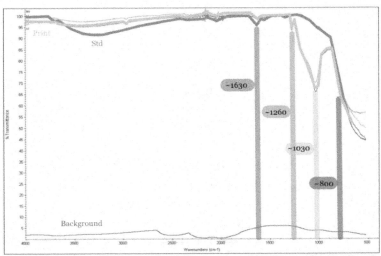

Figure 3. FTIR spectrum of TiO_2 printed and standard (*sans*-PDMS) sample. The lines have been highlighted for greater clarity. The line marked *Print* is the averaged spectra for a printed sample. The highlighted line marked, *Std*, is a standard, non-printed sample. The line marked, *Background,* is in fact the averaged background. Both the standard and printed sample spectra are background subtracted. The vertical axis is intensity, horizontal axis is wavenumber (cm^{-1}).

CONCLUSION

It seems quite plausible that the printing process combined with the elevated temperature brought on by sintering leads to the formation of a contamination layer of SiO_2 and/or some form of a TiO_2-silicate. The SiO_2 layers are likely few and monolayer-like, as the independent peaks that represent two different states of oxygen are completely resolved. The contamination process could form through a scheme of the following: upon removal of the stamp from the gelled TiO_2 pattern, the TiO_2 surface atoms preferentially bind or adhere to the outer-most layer of the PDMS stamp, leaving behind several monolayers of PDMS stamp material. The sintering phase of the TiO_2 sample preparation then easily carbonizes the hydrocarbons present in the contamination layer. This leaves a layer(s) of Si-O that would then readily oxidize into a SiO_2 state. This progression of events may also speak to the increased prevalence of carbon in the printed XPS survey spectrum. Further, the existence of optical phonon modes, which are consistent with those of SiO_2, are present in the FTIR spectra of a printed sample. These modes do not exist in the spectra of a standard sample that was prepared without the utilization of a PDMS stamp.

ACKNOWLEDGMENTS

We gratefully acknowledge the use of the facilities within the LeRoy Eyring Center for Solid State Science at Arizona State University and the facilities within the BioDesign Institute at Arizona State University.

REFERENCES

1. H. Schift, "Nanoimprint Lithography: An Old Story in Modern Times? A Review." *Journal of Vacuum Science & Technology B: Microelectronics and Nanometer Structures* **26.2**, 458-80 (2008).
2. L.J. Guo, "Nanoimprint Lithography: Methods and Material Requirements." *Advanced Materials* **19.4**, 495-513 (2007).
3. D. Qin, X. Younan, and G.M. Whitesides, "Soft Lithography for Micro- and Nanoscale Patterning." *Nature Protocols* **5.3**, 491-502 (2010).
4. D. Xia, Y.B. Jiang, X. He, and S.R.J. Brueck. "Titania Nanostructure Arrays from Lithographically Defined Templates." *Applied Physics Letters* **97.22**, 223106 (2010).
5. J. Zaumseil *et al.*, "Three-Dimensional and Multilayer Nanostructures Formed by Nanotransfer Printing." *Nano Letters* **3.9**, 1223-7 (2003).
6. D. Weiss *et al.*, "Nanoimprinting for Diffractive Light Trapping in Solar Cells." *Journal of Vacuum Science & Technology. B* **28.6** C6M98-103 (2010).
7. D.A. Richmond, Q. Zhang, G. Cao, and D.N. Weiss. "Pressureless Nanoimprinting of Anatase TiO2 Precursor Films." *Journal of Vacuum Science and Technology B* **29.2**, 022603 (2011).
8. Q. Fang *et al.*, "FTIR and XPS investigation of Er-doped SiO_2-TiO films." *Materials Science and Engineering B.* **105**, 209-13 (2003).
9. R.P. Netterfield, P.J. Martin, C.G. Pacey, and W.G. Saintly, "Ion-assisted deposition of mixed TiO_2-SiO_2 films." *J. Appl. Phys.* **66**, 1805-9 (1989).
10. U. Sirimahachai, N. Ndiege, R. Chandrasekharan, S. Wongnawa, and M.A. Shannon, "Nanosized TiO_2 particles decorated on SiO_2 spheres (TiO_2/SiO_2) synthesis and photocatalytic activities." *J. Sol-Gel. Sci. Technol.* **56**, 53-60 (2010).
11. Y. Masuda, Y. Jinbo, T. Yonezawa, and K. Koumoto, "Templated Site-Selective Deposition of Titanium Dioxide on Self-Assembled Monolayers." *Chem. Mater.* **14**, 1236-41 (2002).
12. D. Huang, Z.D. Xiao, J.H. Gu, N.P. Huang, and C.W. Yuan, "TiO_2 thin films formation on industrial glass through self-assembly processing." *Thin Solid Films.* **305**, 110-5 (1996).
13. C.C. Fulton, G. Lucovsky, and R.J. Nemanich, "Electronic states at the interface of Ti-Si oxide on Si(100)." *J. Vac. Sci. Technol. B.* **20,4**, 1726-31 (2002).
14. S.E. Lappi, B. Smith, and S. Franzen, "Infrared spectra of $H_2^{16}O$,$H_2^{18}O$, and D_2O in the liquid phase by single-pass attenuated total internal reflection spectroscopy." *Spectrochemica Acta A.* **60**, 2611-9 (2004).
15. C. T. Kirk, "Quantitative analysis of the effect of disorder-induced mode coupling on infrared absorption of silica." *Phys. Rev. B.* **38,2**, 1255-73 (1988).
16. J. D. Ferguson, E.R. Smith, A.W. Weimer, and S.M. George, "ALD of SiO_2 at Room Temperature using TEOS and H_2O with NH_3 as the Catlayst." *J. Electrochem. Soc.* **151**, G528-35 (2004).
17. C.H. Bjorkman, T. Yamazaki, S. Miyazaki, and M. Hirose, "Analysis of infrared attenuated total reflection spectra from thin SiO_2 films on Si." *J. Appl. Phys.*, **77.1**, 313-7 (1995).

Mater. Res. Soc. Symp. Proc. Vol. 1675 © 2014 Materials Research Society
DOI: 10.1557/opl.2014.862

Electronic structures and optical properties of cuprous oxide and hydroxide

Yunguo Li, Cláudio M. Lousada and Pavel A. Korzhavyi
Division of Materials technology
Department of Materials and Engineering, Royal Institute of Technology (KTH),
SE-100 44 Stockholm, Sweden

ABSTRACT

The broad range of applications of copper, including areas such as electronics, fuel cells, and spent nuclear fuel disposal, require accurate description of the physical and chemical properties of copper compounds. Within some of these applications, cuprous hydroxide is a compound whose relevance has been recently discovered. Its existence in the solid-state form was recently reported. Experimental determination of its physical-chemical properties is challenging due to its instability and poop crystallinity. Within the framework of density functional theory calculations (DFT), we investigated the nature of bonding, electronic spectra, and optical properties of the cuprous oxide and cuprous hydroxide. It is found that the hybrid functional PBE0 can accurately describe the electronic structure and optical properties of these two copper(I) compounds. The calculated properties of cuprous oxide are in good agreement with the experimental data and other theoretical results. The structure of cuprous hydroxide can be deduced from that of cuprous oxide by substituting half Cu^+ in Cu_2O lattice with protons. Compared to Cu_2O, the presence of hydrogen in CuOH has little effect on the ionic nature of Cu–O bonding, but lowers the energy levels of the occupied states. Thus, CuOH is calculated to have a wider indirect band gap of 2.73 eV compared with the Cu_2O band gap of 2.17 eV.

INTRODUCTION

Copper has been used for thousands of years for a variety of purposes most often as pure metal [1,2]. At present, it is the most suitable candidate material for the longtime storage of high-level radioactive spent nuclear waste in a deep underground repository [3,4]. Such application requires accurate knowledge of the behavior of copper under deep geological conditions. Among the copper compounds containing oxygen and hydrogen, the monovalent copper (cuprous) compounds [5,6] are less stable than the divalent compounds (in terms of formation energy), but are still very important in corrosion processes or in applications such as transparent electronics.

Cuprous oxide Cu_2O has been extensively studied and is well known for its prototypical p-type conducting behavior [7-11]. The cuprite crystal structure of Cu_2O contains two sublattices: one Cu^+ face-centered cubic and one O^{2-} body-centered cubic. The conducting behavior has been attributed to the existence of Cu vacancies, V_{Cu} [11-14]. It has been reported that the Cu vacancies are the energetically favored sites for the incorporation of hydrogen impurities [9], which implies a relation between Cu_2O and cuprous hydroxide CuOH. A crystalline form of cuprous hydroxide was recently studied with theoretical and experimental techniques [5,6]. Its atomic structure resembles that of Cu_2O but with half of the Cu^+ replaced by protons. The two protons bound to each O^{2-} lay at different distances from the O^{2-} anion: one is closer (0.973 Å) while the other is further away (2.204 Å). It is therefore interesting to investigate in which way cuprous oxide Cu_2O and cuprous hydroxide CuOH are related by their physical and chemical properties. In the present work, we use *ab initio* electronic structure calculations and a hybrid functional approach to investigate the electronic structure, nature of bonding, and optical properties of the cuprous oxide and cuprous hydroxide.

THEORY

The calculations using the Vienna *ab-initio* simulation package (VASP) [15,16] were based on density functional theory (DFT) and employed the projector augmented wave (PAW) method as implemented by Kresse and Joubert [17,18]. A plane-wave basis set with a cutoff of 800 eV was chosen for all calculations. At this cutoff energy the relative energies were converged to 1×10^{-6} eV. All the considered structures were fully relaxed. The hybrid functional PBE0 [19]—which contains 25% of the exact nonlocal Hatree-Fock (HF) exchange, 75% of the PBE exchange, and 100% of the PBE correlation energy—was used. For all calculations a mesh corresponding to $5 \times 5 \times 5$ k-points in the Brillouin zone was used.

The optical properties of materials can be described by the dynamic dielectric function: $\varepsilon(\omega) = \varepsilon_1(\omega) + i\varepsilon_2(\omega)$, where $\varepsilon(\omega)$ is the frequency dependent dielectric function, $\varepsilon_1(\omega)$ is the real part and $\varepsilon_2(\omega)$ is the imaginary part. The imaginary part can be calculated in momentum space by determination of the matrix elements corresponding to occupied and unoccupied electronic states, which were obtained from the self-consistent band structure calculations within the PAW method. The Kramers–Kroning transformation was used to compute the real part. For a detailed description, we refer the reader to the respective literature Ref. [20]. The electron energy-loss function can be derived from the dynamic dielectric function by $L(\omega)=\text{Im}(-1/\varepsilon(\omega))$, where $L(\omega)$ is the electron energy-loss function and Im means the imaginary part. The electron energy-loss function describes the energy loss when electrons pass through a dielectric medium. The complex refractive index N is related to the complex dielectric function by $N^2(\omega)=\varepsilon(\omega)$. The complex refractive index is defined as $N(\omega)=n(\omega)+ik(\omega)$, and its imaginary part is $k(\omega)=\varepsilon_2(\omega)/2n(\omega)$ is related to the absorption coefficient by $\eta=2k(\omega)\cdot\omega/c$. The imaginary part of the complex refractive index represents the attenuation, while the real part accounts for refraction. The reflectivity coefficient R is expressed by the complex refractive index as $R(\omega)=2|(1-N(\omega))/(1+N(\omega))|$.

DISCUSSION
Electronic structure

Cuprous oxide is a semiconductor with prevailingly ionic bonding, as can be seen in its electron localization function (ELF) shown in Figure 1(a). The grade of the ELF is encoded using a color scheme in which values corresponding to higher probabilities of finding like-spin electrons are shown in red while lower probability regions are shown in blue. At the Cu cation sites the ELF has very low values; whereas at the sites of O anions, spherical regions of high probability (0.85) can be seen. Such large difference in the ELF is a signature of the strong ionic character of Cu–O bonding. Besides, there is a weak halo of electron localization function around Cu^+ in the plane perpendicular to Cu-O bond. Such feature is due to $3d$-shell depletion and s-d hybridization of Cu^+, which leads to a weak interaction between Cu cations that stabilises the structure. The electronic density of states (DOS) is characteristic of this type of bonding. As Figure 2(a) shows, there is a clear gap between the Cu-d and O-p states. The hybridization between these two states is very weak. The PBE0-calculated band gap of cuprous oxide is 2.17

eV, in good agreement with experimental data (2.0-2.2 eV) [7,21]. The upper valence band is dominated by Cu-3d states, with two peaks centered at about −1.0 eV and −2.5 eV. The two O-2p peaks are centered at −6.5 eV and −7.5 eV. The lower valence band has one sharp O-2s peak at −21.3 eV. The PBE0-calculated lattice parameters and band gaps are summarized in Table 1.

Table 1. Calculated crystal structure data and band gap for cuprous oxide and hydroxide.

Material	Space group	Lattice parameters, Å		Band gap (PBE0)
		PBE	PBE0	
Cu_2O	$Pn\bar{3}m$	$a = 4.31$	$a = 4.30$	2.17 (direct)
CuOH	$Cm2a$	$a = 5.70, b = 5.30, c = 4.21$	$a = 5.73, b = 5.33, c = 4.14$	2.73 (indirect)

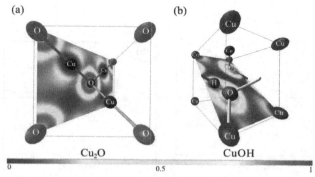

Figure 1. Crystal structures of Cu_2O and CuOH showing cross-sections of electron localization function.

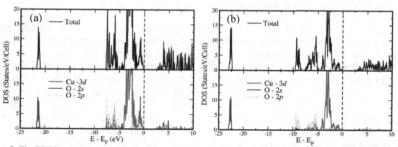

Figure 2. The PBE0-calculated density of electron states of (a) Cu_2O and (b) CuOH. The energy is relative to VBM.

The structure of cuprous hydroxide CuOH, also known as "cuprice", bears great similarity with those of Cu_2O and high-pressure cubic ice [5]. Cuprous hydroxide may be derived from cuprous oxide by replacing half of the Cu^+ in the Cu_2O lattice with protons, so that each O is tetrahedrally coordinated to two Cu^+ and two protons, keeping one proton closer to, and the other further from, the O anion. The positioning of protons in CuOH allows for many different configurations of similar energy as all these structural variants have identical number of chemical bonds. Here we present the results calculated for the smallest unit cell of CuOH shown in Figure 1(b). The ELF shown in the Figure suggests a covalent character of bonding between O and the closest H, while the Cu sites are still characterized by low values of ELF. Around Cu sites there is no lobe protruding towards O, suggesting that the Cu–O bond is still ionic in CuOH. However, chemical bonding to H noticeably changes the O-$2p$ DOS. As seen in Figure 2(b), the O-p DOS in the upper valence band is split into two peaks, centered at about −6.5 eV and −9.0 eV. The peak around −6.5 eV is due to O-p_x, O-p_y and O-p_z, while the other is due to O-p_x and O-p_y states. The O–H bonding can be speculated to be a hybrid of H-s and O-p_z states, which form the bonding and antibonding states. The antibonding peak contains some contribution from Cu states. The hybridization with H-s states lowers the occupied states and also widens the band gap. As can be seen from Table 1, the band gap of CuOH is larger than that of Cu_2O.

Optical properties

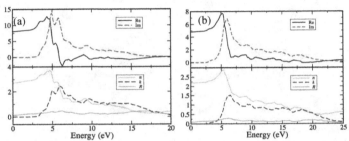

Figure 3. Optical properties of cuprous oxide (a) and hydroxide (b) calculated using the PBE0 functional. Upper panels show the real part (Re) and imaginary (Im) part of dynamic dielectric function. Lower panels show the real part n and imaginary part k of refractive function, as well as the reflectivity coefficient R.

As mentioned above, the PBE0 functional can accurately describe the electronic structure of the two compounds. Therefore, we use the calculated electronic spectra to derive the optical properties. Figure 3(a) shows the complex dielectric function, complex refractive index, and reflectivity coefficient for cuprous oxide. The upper panel shows the real and imaginary parts of the dielectric function. The static dielectric constant $\varepsilon_1(0)$ of Cu_2O is calculated to be 8.1, which falls into the upper end of the experimental value 7.6±0.4 [10]. The imaginary part $\varepsilon_2(\omega)$ is smooth at low energies and reaches a maximum at 5.0 eV, with a shallow shoulder to the left. Although cuprous oxide is a direct band gap semiconductor, the dipole transition matrix element is zero between the valence band maximum (VBM) and conduction band minimum (CBM). So the optical band gap derived by extrapolation from the $\varepsilon_2(\omega)$ curve, 2.8 eV, is larger than the band gap. The highest peak corresponds to the transition between CBM and Cu-$3d$ peaks below the VBM, as can be seen from the DOS of Figure 2. The peak at 5.8 eV in the $\varepsilon_2(\omega)$ curve also corresponds to the transition from Cu-$3d$ states to CBM. Two more peaks located at 7.2 eV and 9.4 eV correspond to the transition from O-$2p$ states to CBM. In the lower panel of Figure 3(a), the maximum of k is 2.6, indicating strong light absorption at around 6.0 eV. The static refractive index is 2.8. The reflectivity coefficient is smooth within the whole energy range.

The complex dielectric function, complex refractive index, and reflectivity coefficient of cuprous hydroxide are shown in Figure 3(b). The real and imaginary parts of the dynamic dielectric function are depicted in the upper panel. The obtained static dielectric constant $\varepsilon_1(0)$ is 4.9 for the hydroxide, which is much lower than for cuprous oxide. This may be due to the fact that replacing Cu with H decreases the polarity of the unit cell. The $\varepsilon_2(\omega)$ has a little plateau before reaching the highest peak at 6.2 eV. This is because cuprous hydroxide is an indirect semiconductor and the direct transition from VBM to CBM requires a contribution from lattice vibrations. The highest peak corresponds to the transition between CBM and Cu-$3d$ peaks lying below Fermi level, similarly to cuprous oxide. The two peaks located at 9.1 and 11.6 eV are due to the transitions from Cu-$3d$ states and O-$2p$ states to CBM, respectively. The optical band gap is equal to the indirect band gap for CuOH. In the lower panel of Figure 3(b), the static refractive index is 2.4. The real and imaginary parts of the static refractive index are smaller than those of cuprous oxide, indicating better light permittivity of cuprous hydroxide. The imaginary part k has a maximum of 1.5, and vanishes at 25 eV. When the energy of the photon is higher than 25 eV, the frequency of incident light exceeds the inherent oscillation frequency of cuprous hydroxide. Beyond this energy, all the quantities responsible for absorption will extinct.

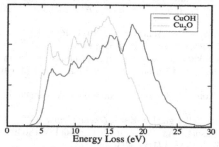

Figure 4. Energy-loss function for Cu_2O and CuOH.

The energy-loss function has been deduced from the dynamic dielectric constant. This function can be compared with experimental spectroscopic data (e.g. EELS, EXELF), which provide information about the interactions of the material with an incident electron beam. The derived energy-loss spectra for both cuprous oxide and hydroxide presented in Figure 4 clearly show the energy ranges corresponding to the electronic excitations of the different orbitals in the low-loss region. The broad spectra appear in the range 2.8-22.5 eV for Cu_2O and 2.9-27.5 eV for CuOH, respectively. The highest peak in the low-loss spectrum is the bulk plasmon peak, corresponding to the collective oscillation frequency of the loosely bound electrons. The calculated plasmon energy is 14.7 eV for cuprite and 18.3 eV for cuprice.

CONCLUSIONS

Based on first-principles DFT calculations, using the hybrid functional PBE0, we have successfully reproduced the electronic and optical properties of cuprous oxide. This method was also applied to cuprous hydroxide—which is derived from Cu_2O by substituting half of the Cu cations with protons. Compared to Cu_2O, the hydrogen in CuOH has little effect on the ionic nature of the Cu-O bonding but lowers the energy levels of the occupied states by giving a covalent character to the O–H bond. Thus CuOH has a widened indirect band gap of 2.73 eV compared to Cu_2O. The similarity in the electronic structure of these compounds is also reflected in their optical properties. Their dynamic dielectric functions are similar, with small differences in the peak positions. From comparison of their optical properties, cuprous hydroxide may be expected to have better light permittivity. The electron energy-loss spectra were also calculated, yielding bulk plasmon energies of 14.7 and 18.3 eV for Cu_2O and CuOH, respectively.

ACKNOWLEDGMENTS

The Swedish Nuclear Fuel and Waste Management Company (SKB) is acknowledged for financial support. Yunguo Li acknowledges support by the Chinese Scholarship Council (CSC). The computations were performed on resources provided by the Swedish National Infrastructure for Computing (SNIC) at the National Supercomputer Center (NSC), Linköping, and at the PDC Center for High-performance Computing, Stockholm.

REFERENCES

1. Y. Lee, J. R. Choi, K. J. Lee, N. E. Stott, D. Kim, *Nanotechnology*, 2008, **19**(41), 415604.
2. A. Atkinson, S. Barnett, R. J. Gorte, J. T. S. Irvine, A. J. McEvoy, M. Mogensen, J. Vohs, *Nature Mater.*, 2004, **3**(1), 17.
3. F. King, C. Padovani, *Corros. Eng. Sci. Technol.*, 2011, **46**(2), 82.
4. B. Rosborg, L. Werme. *J. Nucl. Mater.*, 2008, **379**(1), 142.
5. P. A. Korzhavyi, I. L. Soroka, E. I. Isaev, C. Lilja, B. Johansson, *PNAS* 2012, **109**, 686.
6. I. L. Soroka, A. Shchukarev, M. Jonsson, N. V. Tarakina, P. A. Korzhavyi, *Dalton Trans.* 2013, **42**(26), 9585.
7. M. Hara, T. Kondo, M. Komoda, S. Ikeda, K. Shinohara, A. Tanaka, J. N. Kondo, *Chem. Comm.* 1998, (3), 357.
8. P. E. DeJongh, D. Vanmaekelbergh, J. J. Kelly, *Chem. Comm.* 1999, (12), 1069.
9. D. O. Scanlon, G. W. Watson, *Phys. Rev. Lett.*, 2011, **106**(18), 186403.
10. E. C. Heltemes, *Phys. Rev.*, 1966, **141**(2), 803.
11. M. Heinemann, B. Eifert, and C. Heiliger, *Phys. Rev.*, *B*, 2013, **87**(11), 115111.
12. H. Raebiger, S. Lany, and A. Zunger, *Phys. Rev. B* 2007, **76**(4), 045209.
13. D. O. Scanlon, B. J. Morgan, and G. W. Watson, *J. Chem. Phys.* 2009, **131**(12), 124703.
14. D. O. Scanlon, B. J. Morgan, G. W. Watson, and A. Walsh, *Phys. Rev. Lett.* 2009, **103**(9), 096405.
15. G. Kresse and D. Joubert, *Phys. Rev. B* **59**(3), 1758 (1999).
16. P. E. Blöchl, *Phys. Rev.*, *B*, 1994, **50**(24), 17953.
17. G. Kresse, J. Furthmüller, 1996, **6**(1), 15.
18. G. Kresse, J. Furthmüller, *Phys. Rev.*, *B*, 1996, **54**(16), 11169.
19. C. Adamo, V. Barone, J. Chem. Phys., 1999, **110**(13), 6158.
20. M. Gajdoš, K. Hummer, G. Kresse, J. Furthmüller, F. Bechstedt, *Phys. Rev.*, *B*, 2006, **73**(4), 045112.
21. S. N. Kale, S. B. Ogale, S. R. Shinde, M. Sahasrabuddhe, V. N. Kulkarni, R. L. Greene, T. Venkatesan. *Appl. Phys. Lett.*, 2003, **82**(13), 2100.

Mater. Res. Soc. Symp. Proc. Vol. 1675 © 2014 Materials Research Society
DOI: 10.1557/opl.2014.863

Cobalt oxide-tungsten oxide nanowire heterostructures: Fabrication and characterization

Nitin Chopra,[1,2,*] Yuan Li,[1] Kuldeep Kumar[1]

[1]Metallurgical and Materials Engineering Department, Center for Materials for Information Technology (MINT), The University of Alabama, Tuscaloosa, AL 35487, U.S.A.
[2]Department of Biological Sciences, The University of Alabama, Tuscaloosa, AL 35487, U.S.A.
*Corresponding Author E mail: nchopra@eng.ua.edu, Tel: 205-348-4153, Fax: 205-348-2164

ABSTRACT

Nanowire heterostructures comprised of cobalt oxide and tungsten oxide were fabricated in a core/shell configuration. This was achieved by sputter coating tungsten oxide shells on standing cobalt oxide nanowires on a substrate. To ensure the polycrystallinity of tungsten oxide shell, the nanowire heterostructures were subjected to post-sputtering annealing process. The cobalt oxide nanowires for this study were grown employing a thermal method via vapor-solid growth mechanism. The crystal structures, morphologies, dimensions, and phases at various growth stages of nanowire heterostructures were studied using high resolution electron microscopy, energy dispersive spectroscopy, and X-ray diffraction methods. The interfaces of these nanowire heterostructures were also studied and showed variation in the lattice spacing across the heterostructure diameter. Results indicated that the cobalt oxide nanowires survived multiple processing steps and resulted in stable heterostructure configurations. The investigation shows, for the first time, a dry processing route for the formation of such novel nanowire heterostructures.

INTRODUCTION

Nanowire heterostructures are novel 1-D nanoarchitectures that can exhibit unique surface functionality, chemical and electrical properties, and light-matter interactions [1,2]. For photocatalysis, the aim is to develop 1-D geometry with direction-dependent light-matter interactions [3,4]. For example, 1-D geometry such as core/shell nanowires allow for greater absorption of light in longitudinal direction and result in efficient charge transport (or charge carrier generation) in radial direction [4,5]. However, this is dependent on energy and crystal structure, morphology, size, and material selection for the core/shell nanowires. In addition, one of the challenges for photoactive nanostructures is to eliminate or mitigate the use of noble metal systems and explore cheaper oxides [6]. In this regard, the oxides of cobalt and tungsten are of interest due to stability of these oxides, ability to result in tunable and size-dependent band gap energies, and can be fabricated through a variety of approaches [4,7,8]. Tungsten oxide is a visible light active catalyst (band gap energy ~2.8 eV) when sacrificial agents or precious metals are present and cobalt oxide is a p-type semiconductor (band gap energy ~1.6 eV) and of potential interest for photocatalysis as well as batteries [4,9,10].

The major goal of this research was to develop novel oxide-based nanowire heterostructures with controlled morphology, chemical composition, and interface. Here, we report the fabrication of cobalt oxide-tungsten oxide nanowire heterostructures. These heterostructures remained standing on the substrate due to the dry and surfactant-free processing route. The morphology, chemical composition, and crystal structure of the heterostructures were characterized by microscopic and spectroscopic techniques.

EXPERIMENT

Materials and methods: Cobalt foil was purchased from Alfa Aesar (Ward Hill, MA), Cobalt powders were purchased from ACROS (Geel, Belgium). (100), n-type silicon wafers were purchased from IWS (Colfax, CA). 68% nitric acid was purchased from VWR (Atlanta, GA). DI water (18.1 MΩ-cm) was obtained using a Barnstead International DI water system (E-pure D4641). All chemicals were used without further purification. Branson 2510 Sonicator (Danbury, CT) was used to assist the surface cleaning of Co foil/powders in acetone and DI-water. Thermal oxidation growth of cobalt oxide nanostructures was carried out in a MTI GSL-1100X CVD furnace. ATC ORION sputtering system (AJA international, Inc., North Scituate, MA) was used. Cobalt targets (99.999%) were provided by AJA International, Inc. 5% O$_2$/Ar gas cylinders were purchased from Airgas South (Tuscaloosa, AL).

Thermal oxidation of cobalt oxide nanostructures: Co foils were sequentially cleaned by sonicating in acid, acetone, and DI-water. These substrates were immediately put in the growth furnace. The growth of Co oxide nanostructures was carried out at 300 ˚C for 5 h in presence of O$_2$/Ar mixture as the carrier gas.

Cobalt oxide-tungsten oxide nanowire heterostructure synthesis: The cobalt oxide nanowire samples were put in the sputtering chamber and sputtered directly with tungsten oxide in presence of Ar and at 100 W and 10 mTorr. The obtained nanowire heterostructures were further annealed at 300 ˚C for 3 h.

Characterization: Scanning Electron Microscopy (SEM) images were obtained using FE-SEM JEOL-7000 equipped with energy dispersed X-ray spectroscopy (EDX). Tecnai F-20 was used to collect Transmission Electron Microscopy (TEM) images at 200 kV. TEM samples were prepared by dispersing as-prepared samples on lacey carbon TEM copper grids purchased from Ted Pella Inc. (Redding, CA).

DISCUSSION

Cobalt oxide nanowires were grown in a thermal method, where the mechanism was governed by vapor-solid mechanism [11]. This process resulted in standing nanowires with diameter of ~76.81 ± 14.37 nm and length of ~5.33 ± 2.23 μm. After the sputter deposition of tungsten oxide onto cobalt oxide nanowires, the resulting nanowire heterostructures were ~88.66 ± 11.39 nm in diameter. The increase in the diameter was around ~12 nm thick tungsten oxide shell. Finally, these nanowire heterostructures were annealed and exhibited diameter of ~102.66 ± 20.63 nm resulting in tungsten oxide thickness of ~25-28 nm. This indicates that post-sputtering annealing process resulted in redistribution of material around nanowires. Based on previous observations of the authors, this annealing process must have involved surface migration of tungsten oxide from the flat substrate region and nanowire base towards the nanowire tip [4,12]. Such surface migration has been attributed to surface tension, capillary forces, and chemical potential differences between the flat substrate and nanowire surface [12].

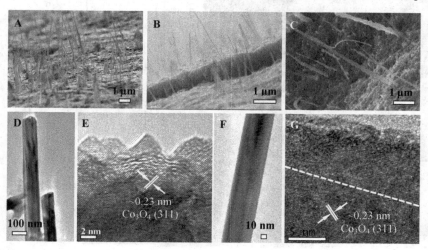

Figure 1. SEM images of (A) cobalt oxide nanowires and cobalt oxide-tungsten oxide nanowire heterostructures (B) before and (C) after annealing process. TEM images of (D and E) cobalt oxide nanowires and (F and G) cobalt oxide-tungsten oxide nanowire heterostructures before annealing.

SEM images show standing cobalt oxide nanowires and cobalt oxide-tungsten oxide nanowire heterostructures before and after the annealing process (Figure 1A-C). Cobalt oxide nanowires exhibited (311) planes of Co_3O_4 (Figure 1D and E). The surface of nanowires was also observed to be rough. The cobalt oxide-tungsten oxide nanowire heterostructures before annealing showed amorphous tungsten oxide shell (Figure 1F and G). This was converted into polycrystalline shell after the annealing process and EDS showed the presence of various elements (W, Co, O, Figure 2A-C). The faceted growth of tungsten oxide after annealing process can be thermodynamically explained on the basis of increased solid aggregate vapor pressure, which resulted in a change in the surface energy to maintain Gibb's free energy at zero [4,12]. Polycrystalline lattice with variations in lattice spacing from pure tungsten oxide to pure cobalt oxide were observed along the nanowire heterostructure diameter (Figure 2D-G). This variation in lattice spacing could be explained by the interdiffusion of materials. The electron diffraction also showed the presence of only three components (Co_3O_4, CoO, and WO_3), where Co_3O_4 may have been present within the thin oxide film on the substrate and by virtue of vapor-solid growth mechanism for nanowires [11].

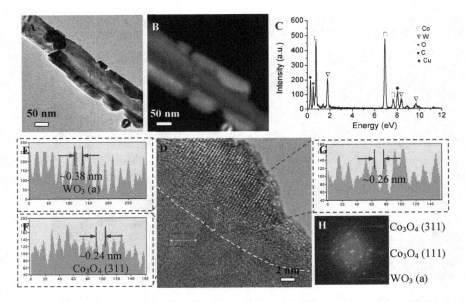

Figure 2. (A) TEM image of the cobalt oxide-tungsten oxide nanowire heterostructures after annealing process and corresponding **(B)** STEM image showing the contrast between the shell and core nanowire. **(C)** EDS spectra of the nanowire heterostructures. **(D)** HRTEM showing the lattice variations in different region of the nanowire heterostructures. **(E-G)** lattice evaluation for various regions of the nanowire heterostructures. **(H)** FFT image (electron diffraction) corresponding to (D).

XRD further confirmed the presence of different phases and elements within the nanowire heterostructures at various stages of fabrication. Various crystallographic planes for CoO, Co_3O_4, and WO_3 are shown in the XRD results (Figure 3 and Table 1). The peaks for the CoO were attributed to the presence of this oxide in the oxide thin film on the cobalt foil substrate. Furthermore, after sputter deposition and annealing of tungsten oxide, minor shifts in Co_3O_4 peaks were observed suggesting that the deposition of tungsten oxide layers led to strains within the nanowire lattice. However, there was not a significant change in XRD peak shifts post-annealing compared to pre-annealing, which suggests that nanowires could sustain multiple processing steps and high temperatures. Further research in the authors' laboratory is currently focused on applications of these nanowire heterostructures.

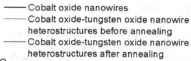

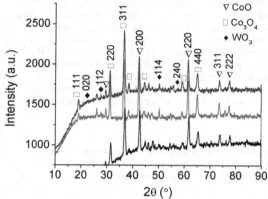

Figure 3. XRD for cobalt oxide nanowires and cobalt oxide-tungsten oxide nanowire heterostructures before and after annealing.

Table 1. Precise XRD peak assignment for various components at different stages of nanowire heterostructure fabrication.

	Crystal indices	Cobalt oxide nanowires	Cobalt oxide-tungsten oxide nanowire heterostructures before annealing	Cobalt oxide-tungsten oxide nanowire heterostructures after annealing
Co_3O_4	(111)	19.50	19.02	18.93
	(220)	31.57	31.16	31.16
	(311)	36.90	36.51	36.61
	(440)	65.40	64.88	64.98
CoO	(200)	42.78	42.36	42.36
	(220)	61.79	61.38	61.38
	(311)	73.92	73.51	73.51
	(222)	77.73	77.40	77.38
WO_3	(020)	/	22.45	22.41
	(112)	/	27.67	27.65
	(114)	/	50.18	50.18
	(240)	/	57.51	57.41

CONCLUSIONS

This study demonstrates for the first time fabrication of standing cobalt oxide-tungsten oxide nanowire heterostructures directly onto a substrate in a surfactant-free and dry fabrication method. The cobalt oxide nanowire growth was achieved using a thermal growth approach by directly utilizing cobalt substrate followed by sputter deposition of tungsten oxide and annealing to result in polycrystalline shells of tungsten oxide around cobalt oxide nanowires. The heterostructures were studied for their crystal structure and lattice spacing across the interface and a variation in lattice spacing was observed from along the heterostructure diameter. XRD further confirmed the presence of CoO (film on the substrate), Co_3O_4 (nanowire), and WO_3 (shell). A minor shift in nanowire XRD peaks after heterostructure formation suggest that the lattice of core nanowire was strained due to the presence of shells and post-growth processing. Such novel nanowire heterostructures can be very useful for device and catalytic applications.

ACKNOWLEDGMENTS

This work was funded by National Science Foundation NSF-EPSCoR RII award and Research Grant Committee awards to Dr. Chopra. The authors thank CAF facilities and MINT Center. The authors thank Dr. S. Kapoor for proof reading the manuscript.

REFERENCES

1. N. Chopra, *Mater. Technol. Adv. Perform. Mater.* **25**, 212 (2010).
2. G. Shen, D. Chen, Y. Bando and D. Golberg, *J.Mater. Sci. Technol.* **24**, 541 (2008).
3. L. Wang, H. Wei, Y. Fan, X. Gu and J. Zhan, *J. Phys. Chem. C* **113**, 14119 (2009).
4. W. Shi and N. Chopra, *ACS Appl. Mater. Interf.* **4**, 5590 (2012).
5. P. Tongying, V. V. Plashnitsa, N. Petchsang, F. Vietmeyer, G. J. Ferraudi, G. Krylova, and M. Kuno, *J. Phys. Chem. Lett.*, **3**, 3234 (2012).
6. F. E. Osterloh, *Chem. Soc. Rev.* **42**, 2294 (2013).
7. M, R, Waller, T. K. Townsend, J. Zhao, E. M. Sabio, R. L. Chamousis, N. D. Browning, and F. E. Osterloh, *Chem. Mater.* **24**, 698 (2012).
8. F. E. Osterloh and B. A. Parkinson, B. A. *MRS bulletin*, **36**, 17 (2011).
9. W. Y. Li, L. N. Xu, and J. Chen, *Adv. Funct. Mater.* **15**, 851 (2005).
10. P. P. González-Borrero, F. Sato, A. N. Medina, M. L. Baesso, A. C. Bento, G. Baldissera, C. Persson, G. A. Niklasson, C. G. Granqvist, and A. F. da Silva, *Appl. Phys. Lett.* **96**, 061909 (2010).
11. B. Varghese, C. H. Teo, Y. Zhu, M. V. Reddy, B.V.R. Chowdari, A.T. S. Wee, V. B. C. Tan, C. T. Lim, and C-H. Sow, *Adv. Funct. Mater.* **17**, 1932 (2007).
12. W. Shi and N. Chopra, *J. Nanopart. Res.* 13, 851 (2011).

Mater. Res. Soc. Symp. Proc. Vol. 1675 © 2014 Materials Research Society
DOI: 10.1557/opl.2014.875

Electrical Properties at Grain Boundaries Influenced by Cr^{3+} Diffusion in $SnO_2.ZnO.Nb_2O_5$-Films Varistor Prepared by Electrophoresis Deposition

Glauco M. M. M. Lustosa[1], João Paulo C. Costa[2], Leinig A. Perazolli[1], Maria A. Zaghete[1].
[1]Instituto de Química - UNESP, Araraquara, SP, 14.800-900, Brazil
[2]Centro Universitário de Araraquara - UNIARA, Araraquara, SP, 14.801-340, Brazil

ABSTRACT

SnO_2-based varistors are strong candidates to replace the ZnO-based varistors due to ordering fewer additives to improve its electrical behavior as well as by showing similar nonlinear characteristics of ZnO varistors. In this work, SnO_2-nanoparticles based-varistors with addition of 1.0 %mol of ZnO and 0.05 %mol of Nb_2O_5 were synthesized by chemical route. $SnO_2.ZnO.Nb_2O_5$-films with 5 μm of thickness were obtained by electrophoretic deposition (EPD) of the nanoparticles on Si/Pt substrate from alcoholic suspension of SnO_2-based powder. The sintering step was carried out in a microwave oven at 1000 °C for 40 minutes. Then, Cr^{3+} ions were deposited on the films surface by EPD after the sintering step. Each sample was submitted to different thermal treatments to improve the varistor behavior by diffusion of ions in the samples. The films showed a nonlinear coefficient (α) greater than 9, breakdown voltage (V_R) around 60 V, low leakage current ($I_F ≈ 10^{-6}$ A), height potential barrier above 0.5 eV and grain boundary resistivity upward of 10^7 Ω.cm.

INTRODUCTION

A varistor is an electrical device-based semiconductor material used for protection against voltage surges the electricity grid and overvoltages in electronic circuits and electricity systems. The SnO_2-based varistors were introduced by Pianaro as an alternative to the ZnO-based commercial varistors, that show nonlinear electrical behavior similar to ZnO-varistors, requiring low concentration of modifying agents that promote densification and varistor characteristics and a simpler microstructure without formation of secondary phases [1,2,3].

The electrical conduction in SnO_2 occurs due to defects such as oxygen vacancies and interstitial tin atoms (Eq. 1 and 2) and defects caused by the presence of impurities. It may also occur by adding other elements (dopants/modifiers) [4] that act as electron acceptor or donor for ceramic matrix.

$$SnO \xrightarrow{SnO_2} Sn^{''}_{Sn} + V_O^{\bullet\bullet} + O^x_o \qquad (1)$$

$$SnO_2 \xrightarrow{SnO_2} Sn_i^{\bullet\bullet\bullet\bullet} + V_{Sn}^{''''} + 2O^x_o \qquad (2)$$

$$MO \xrightarrow{SnO_2} M^{''}_{Sn} + V_O^{\bullet\bullet} + O^x_o \qquad (3)$$

The defects of $M^{''}_{Sn}$ (Eq. 3) trap electrons from the others modifiers and create a potential barrier in the grain boundary region. The electrical conductivity of the varistor system can be improved with the addition of pentavalent ions [1,4,5].

In this work, studies were conducted on obtaining SnO_2-films doped with ZnO and Nb_2O_5 (added to the system to improve the densification and the grain conductivity, respectively) aiming at varistor properties. Cr^{3+} ions were added on the surfaces of the sintered samples in order to direct modification of the potential barrier formed in grain boundary region and improve the electrical behavior.

EXPERIMENTAL DETAILS

SnO$_2$, ZnO and Nb$_2$O$_5$ polymeric solutions were prepared using the Pechini method [6]. Stoichiometric amounts of each polymeric solution were mixed with stirring and heating at 90 °C and then pre-calcined at 350 °C/2 hours in a muffle furnace. After calcination, the material was milled at 500 rpm during 1 hour in a ball mill. Then, the powder was calcined at 500 °C for 2 hours in the muffle furnace. The particles of ceramic powder with composition a (98.95)SnO$_2$.(1.00)ZnO.(0.05)Nb$_2$O$_5$ were used to prepare an ethylic suspension and then were separated by gravimetry using stokes law.

The smaller particles of similar sizes were used to obtain films by electrophoretic deposition (EPD). The SnO$_2$-based powder, kept in an ethylic suspension (ratio of 0.014 g of powder/20 ml of ethanol), was taken to EPD for deposition of SnO$_2$-based particles on the Si(100)/TiO2(10.000 A°)/Ti(200 A°)/Pt(1500 A°) substrate, with applying a voltage of 2 kV for 10 minutes. The electrodes of the system were modified with magnets, and 0.02 g of iodine was added into the ethylic suspension to increase the rate of deposition of the particles (after the films were obtained, the iodine was eliminated from the samples with heat treatment at 250 °C/30 min). Then, the films were sintered at 1000 °C for 40 minutes in a microwave oven. After the sintering step, a layer of Cr^{3+} ions was deposited on the films' surface (also by EPD, using voltage of 2 kV for 5 minutes). Finally each sample was subjected to different thermal treatments in microwave to promote the Cr^{3+} diffusion in the material, as shown in Table 1.

The samples Film B and Film C were performed in triplicate to verify the reproducibility of the applied methodology in the electrical properties, generating Films B1/B2 and Films C1/C2 which were submitted to the same conditions as Film B and Film C, respectively.

Table I. Conditions of the samples subjected to heat treatment applied in microwave oven.

Sample	Sintering	Deposition of Cr^{3+}	Heat Treatment for Cr^{3+} diffusion
Film 0	1000 °C/40min	No	-----
Film A	1000 °C/40min	Yes	1000 °C/5 min
Film B	1000 °C/40min	Yes	1000 °C/10 min
Film C	1000 °C/40min	Yes	1000 °C/15 min

After heat treatments of the samples, platinum upper electrodes were deposited on the films' surface by *RF Sputtering*, and then the films were electrically characterized. The films were subjected to characterization I *vs* V as a function of temperature and impedance spectroscopy. For I *vs* V analysis, the Eq. 4 was used to check the nonlinear coefficient of curves [7,8].

$$\alpha = (\log E_2 - \log E_1)^{-1} \qquad (4)$$

where E$_1$ is the electrical field when current density is 1 mA/cm^2 and E$_2$ is the electrical field when current density is 10 mA/cm^2.

The analysis of the linear region of the low current (0 mA/cm² to 1 mA/cm²) of I *vs* V graphs of samples subjected to electrical characterization as a function of temperature was performed to evaluate the resistivity (ρ_{GB}) shown in the grain boundary. The conduction of SnO$_2$-based varistor was studied by Pianaro [9] and determined as being Schottky type, ie, the conduction occurs through the potential barrier formed at the grain boundary region from the action of the electric field and temperature. Equation 5 describes this behavior: [10,11]

$$J_S = A^* . T^2 . \exp[-(\phi_b - \beta E_{1/2})/kT] \qquad (5)$$

where A* is Richardson's constant, ϕ_b is height of the potential barrier, E represents the electric field, T is the temperature in Kelvin and β a parameter related to the width of the potential barrier.

Impedance spectroscopy measurements were obtained to evaluate Cr^{3+} diffusion in the grain with the application of a frequency of 10 Hz to 1 MHz (Autolab). With the data collected from the curves, it was possible to verify the grain resistivity ρ_G ($\Omega.cm$) in Eq. 6: [12]

$$\rho_G = R(S/l) \qquad (6)$$

where R is the resistance (Ω) obtained from the diameter of the semicircle of high-frequency, S (cm^2) being the electrode area and l (cm) the thickness of the sample.

DISCUSSION

Studies were conducted to obtain the best sintering condition of the samples when temperatures at 900 and 1000 °C were used for sintering step. According to the limitations of the microwave oven used, the best condition was observed when applied to 1000 °C for 40 minutes. Figure 1 shows SEM images of a sintered sample. It was possible to observe that there was growth of particles, formation of necks between the particles and also that the deposition by EPD allowed the films to obtain a regular thickness (5 μm) and the presence of pores.

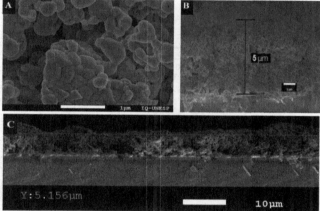

Figure 1. SEM of the film obtained by electrophoresis and sintered at 1000 °C/40 min in a microwave oven: view of the surface (A), view of cross section (B and C).

A layer of Cr^{3+} ions was deposited on the surface of the sintered films also using EPD technique. The application of chromium on the surface of the film was chosen as the method after a few previously published studies about SnO_2 varistors [13] had suggested that chromium segregates in the grain boundary region when high temperature is used for sintering/densification of the varistor, improving the properties in the grain boundary region (resistivity, nonlinear coefficient).

From the data of I vs V as a function of temperature, the grain boundary resistivity of the samples was determined. As shown in Figure 2, it was observed that the resistor characteristic decreases with the increase of temperature. Curves of $\ln(J/T^2)$ vs $E_{1/2}$ were plotted from the I vs V analysis as a function of temperature, making extrapolation to values of E = 0 to be possible to obtain the values of $\ln(J_0/T^2)$ from the intersection of the curves with the axis. The values of $\ln(J_0/T^2)$ were used to construct a function of 1/T, allowing for the calculation of the value of the height (φ_b) of the potential barrier formed at the grain boundary region. Table 2 shows the calculated values for the varistor properties, such as the nonlinear coefficient (α), breakdown voltage (V_R), leakage current (I_F), resistivity of the grain boundary (ρ_{GB}) and height of the potential barrier (ϕ_b). The V_R is obtained when the varistor begins to show the electrical conductivity, determined as the value of applied voltage when current density is 1 mA/cm^2. The I_F was determined as the value of the current when the voltage reached 70 % of the V_R. The I_F represents the current that passes through the material before it attain the V_R.

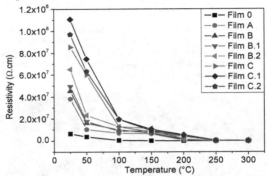

Figure 2. Behavior of the resistivity after the **I** x **V** analysis as a function of temperature for SnO$_2$-based film sintered at 1000 °C/40 min and thermal treated at 1000 °C to Cr^{3+} diffusion.

Table II. The calculated values of nonlinearity coefficient (α), breakdown voltage (V_R), leakage current (I_F), resistivity of the grain boundary (ρ_{GB}) and height of the potential barrier (ϕ_b).

SnO-based thick film	Deposition Cr^{3+}	Heat treatment to Cr^{3+} diffusion	α	V_R (Volts)	I_F (A)	ρ at 298 K (Ω.cm)	ϕ (eV)
Film 0	No	-----	1.7	5.0	6.8x10^{-4}	6.5x10^{-6}	0.34
Film A	Yes	1000 °C/5 min	9.6	24.0	4.6x10^{-4}	3.8x10^{-7}	0.47
Film B			10.3	62.5	4.4x10^{-5}	4.8x10^{-7}	0.56
Film B.1	Yes	1000 °C/10 min	10.5	70.4	6.2x10^{-6}	4.9x10^{-7}	0.58
Film B.2			10.8	78.5	1.6x10^{-6}	6.5x10^{-7}	0.60
Film C			11.4	59.5	4.8x10^{-6}	8.6x10^{-7}	0.61
Film C.1	Yes	1000 °C/15 min	13.8	75.5	8.5x10^{-6}	1.1x10^{-8}	0.68
Film C.2			11.5	60.4	2.5x10^{-6}	9.7x10^{-8}	0.60

It was observed that the addition and promotion of Cr^{3+} diffusion by increasing the heat treatment precipitates a significant change in the electrical properties, thus resulting in an increase of the values of α (from 1.7 to over 9) and E_R (from 2.5 Volts to over 50 Volts) and a decrease of I_F (from 10^{-4} to less than 10^{-6} A). This behavior can well be explained due to the defects which are created by the additives that are responsible for the modification of potential barriers in the grain boundaries region.

The equivalent circuit model, for varistor ceramics, is composed of two series circuits of a resistance and capacitor in parallel. The Nyquist diagram is composed of two or three semicircles, thus, the electrical response (in terms of impedance) can be modeled as an equivalent circuit or a combination of circuits [14]. As shown in Fig. 3, the samples exhibit only one semicircle, therefore not distinguishing the specific contributions of grain, grain boundary and electrode in the total resistance of the samples. This can be explained by the distribution of defects at the grain boundary and also indicates that the equivalent circuit, for this work, cannot be viewed traditionally.

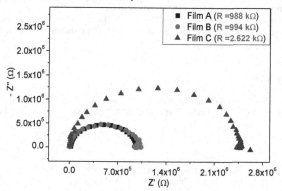

Figure 3. Nyquist diagram for films subjected to thermal treatment at 1000 °C after the deposition of Cr^{3+}. Films sintered at 1000 °C/40 minutes.

The grain has an effect on the varistor characteristic. The impedance analysis (Nyquist diagram) shows the real (Z') vs imaginary (Z'') part of resistance for the samples. Eq.6 was used for determining the resistivity grain with the values shown in Table 3. It is worth observing an increase in electrical resistance of the material when the samples are processed at 1000 °C for 15 minutes after Cr^{3+} deposition. This effect may be explained by the incorporation of chromium into SnO_2 crystalline lattice. It is also worth noting that, for all samples, the grain boundary resistivity (Table 2) is greater than the grain resistivity indicating that the potential barrier is effective in blocking the passage of electric current in the material (until it reaches the breakdown voltage).

Table III. Values calculated of grain resistance (ρ_G).

SnO-based thick film	Deposition of Cr^{3+}	Heat treatment to Cr^{3+} diffusion	Grain Resistence (kΩ.cm)
Film A		1000 °C/5 min	44.9
Film B	Yes	1000 °C/10 min	44.7
Film C		1000 °C/15 min	1180.1

CONCLUSION

- There was good coverage on the substrate by the electrophoretic deposition technique of SnO_2-based nanoparticles, with relative porosity when sintered at 1000 °C/40 min.
- The film presents a varistor behavior with electrical properties that are influenced by the diffusion of Cr^{3+} ions as a function of heat treatment.

- The electrophoretic deposition technique showed good potential for preparation of thick films to be used as low voltage varistors.
- The resistivity values of the grain and boundary grain regions and of the non-linearity coefficient of films increase according to Cr^{3+} diffusion promoted by increasing heat treatment, suggesting the possibility of controlling these parameters by diffusion of Cr^{3+}.
- The values of the varistor parameters for films B1/B2 and C1/C2 demonstrate that the process used to prepare films with varistor properties from the diffusion of Cr^{3+} produced samples with reproducible properties.

ACKNOWLEDGMENTS

The authors thanks to the LMA-IQ for providing the FEG-SEM facilities and the Brazilian research funding agencies CNPq and CEPID/CDMF- FAPESP 2013/07296-2 for providing the financial support of this research project.

REFERENCES

1. A. B. Glot, A. V. Gaponov, A. P. Sandoval-Gracía, *Physica B* **405** 705 (2010).
2. Y. J. Wang, J. F. Wang, C. P. Li, H. C. Chen, W. B. Su, W. L. Zhong, P. L. Zhang, L. Y. Zhao, *J. Mater. Sci. Lett.* **20** 19 (2001).
3. P.A. Santos, S. Maruchin, G. F. Menegoto, A. J. Sara, S. A. Pianaro, *Mater. Lett.* **60** 1554 (2006).
4. M. Cilense, M. A. Ramirez, C. R. Foschini, D. R. Leite, A. Z. Simões, W. Bassi, E. Longo, J. A. Varela, *J. Am. Ceram. Soc.* **96** 524 (2013).
5. R. Metz, D. Koumeir, J. Morel, J. Pansiot, M. Houabes, M. Hassanzadeh, *J. Eur. Ceram. Soc.* **28** 829 (2008).
6. B. D. Stojanivic, C. R. Foschini, M. Cilense, M. A. Zaghete, A. A. Cavalheiro, C. O. Paiva-Santos, E. Longo, J, A Varela,n*Mater Chem Phys* **68** 136 (2001).
7. J. He, Z. Peng, Z. Fu, Z. Wang, X. Fu, *J. Alloy Compd.* **528** 79 (2012).
8. H. Feng, Z. Peng, X. Fu, Z. Fu, C. Wang, L, Qi, H. Miao, *J. Alloy Compd.* **509** 7175 (2011).
9. S. A. Pianaro, P. R. Bueno, P. Olivi, E. Longo, J. A. Varela, *J. Mater. Sci.-Mater. El.* **9** 159 (1998).
10. J. F. Wang, W. B. Su, H. C. Chen, W. X. Wang, G. Z. Zang, *J. Am. Ceram. Soc.* **88** 331 (2005).
11. A. A. Felix, M. O. Orlandi, J. A. Varela, *Solid State Commun.* **151** 1377 (2011).
12. S. K. Tadokoro, E. N. S. Muccillo, *Cerâmica* **47** 100 (2001).
13. M. R. Cássia-Santos, V. C. Sousa, M. M. Oliveira, P. R. Bueno, W. K. Bacelar, M. O. Orlandi, C. M. Barrado, J. W. Gomes, E. Longo, E. R. Leite, J. A. Varela, *Cerâmica* **47** 136 (2001).
14. G. Z. Zang, L. B. Li, H. H. Liu, X. F. Wang, Z. G. Gai, *J. Alloy Compd.* **580** 611 (2013).

Mater. Res. Soc. Symp. Proc. Vol. 1675 © 2014 Materials Research Society
DOI: 10.1557/opl.2014.866

Sol-Gel Synthesis of Nanocrystalline Ni-Ferrite and Co-Ferrite Redox Materials for Thermochemical Production of Solar Fuels

Rahul R. Bhosale[1], Ivo Alxneit[2], Leo L. P. van den Broeke[1], Anand Kumar[1], Mehak Jilani[1], Shahd Samir Gharbia[1], Jamila Folady[1], Dareen Zuhir Dardor[1]

[1] Department of Chemical Engineering, Qatar University, Doha, Qatar.
[2] Solar Technology Laboratory, Paul Scherrer Institute, CH-5232 Villigen PSI, Switzerland.

ABSTRACT

In this contribution, we report the synthesis and characterization of $Ni_xFe_{3-x}O_4$ and $Co_xFe_{3-x}O_4$ redox nanomaterials using sol-gel method. These materials will be used to produce solar fuels such as H_2 or syngas from H_2O and/or CO_2 via solar thermochemical cycles (STCs). For the sol-gel synthesis of ferrites, the Ni, Co, Fe precursor salts were dissolved in ethanol and propylene oxide (PO) was added dropwise to the well mixed solution as a gelation agent to achieve gel formation. Freshly synthesized gels were aged, dried, and calcined by heating them to 600°C in air. The calcined powders were characterized by powder x-ray diffractometer (XRD), BET surface area, as well as scanning (SEM) and transmission (TEM) electron microscopy. Their suitability to be used in STCs for the production of solar fuels was assessed by performing several reduction/re-oxidation cycles using a thermogravimetric analyzer (TGA).

INTRODUCTION

The world's current energy economy is still based to a large extent on the abundance of fossil fuels. This has led to a rapid depletion of the easily accessible oil reserves resulting in continuously rising oil prices. Furthermore severe environmental problems caused by the CO_2 induced greenhouse effect have begun to become apparent [1-4]. Thus, there is a pressing need to develop technologies to produce carbon free renewable fuels such as H_2 or renewable precursors for fuels such as syngas (a mixture of H_2 and CO). The latter can be processed to liquid fuels (gasoline, jet fuel) via the Fischer-Tropsch process [5]. Solar radiation is an essentially inexhaustible energy source that delivers about 100,000 TW to the earth. To harvest the solar radiation and to convert it effectively into renewable fuels from H_2O and captured CO_2 provides a promising path for a future sustainable energy economy.

One of the potential routes to produce solar fuels is metal oxide based solar thermochemical cycles (STCs) [6]. In these cycles, the first step consists of the endothermic thermal reduction of a metal oxide at elevated temperatures releasing O_2. This step is achieved by using concentrated solar radiation as an energy source. The second step consists of the slightly exothermic re-oxidation of the reduced metal oxide at lower temperatures by H_2O, CO_2, or a mixture of the two producing H_2, CO or syngas. Among the many metal oxides investigated so far for solar fuel production, in recent years, research has been focused towards non-volatile mixed metal oxides such as ferrites [7-11]. Ferrites are particularly attractive because their reduced form is a solid and hence the difficult separation of a gaseous metal (or metal oxide) and O_2, a necessary step in cycles based on ZnO/Zn or SnO_2/SnO/Sn, can be eliminated. Previously, ferrites were extensively used for solar thermochemical H_2 production [7-10], while their utilization for CO or syngas production via STCs so far is limited [11-13].

Recently, Bhosale et al. [14-18] synthesized several ferrite materials via sol-gel method and investigated their redox capabilities with their application in solar thermochemical H_2O splitting cycles in mind. The results reported indicate that the ferrites derived via sol-gel method are capable of producing higher amounts of H_2 during multiple thermochemical cycles when compared with ferrites synthesized via different synthesis approaches. Although, the sol-gel derived ferrites were observed to offer promising results in the solar thermochemical H_2 generation, their application for the production of CO or syngas has not yet been investigated.

In this study, we report the synthesis of $Ni_xFe_{3-x}O_4$ and $Co_xFe_{3-x}O_4$ (where, $x = 0.2$ to 1) nanoparticles using sol-gel approach. As-synthesized ferrite powders were characterized by powder x-ray diffractometer (XRD), BET surface area, as well as scanning (SEM) and transmission (TEM) electron microscopy (TEM), and inductively coupled plasma spectrometer (ICP). Derived ferrites were examined towards their thermal reduction and CO_2 splitting ability by performing successive thermochemical cycles using a thermogravimetric analyzer (TGA).

EXPERIMENTAL

Synthesis of ferrites via sol-gel method

The nanostructured $Ni_xFe_{3-x}O_4$ and $Co_xFe_{3-x}O_4$ redox materials were synthesized via sol-gel method and a typical synthesis route is shown in Figure 1a. For the synthesis of $Co_xFe_{3-x}O_4$, the nitrate salts of Co and Fe were added to ethanol and this mixture was sonicated to dissolve the metal salts in the solvent. To this solution obtained after sonication, predetermined amount of propylene oxide (PO) was added dropwise and the gel formation was achieved. The gels obtained were aged for 24 h, dried at 100°C, and then heated upto 600°C and cooled down rapidly in air. Various combinations of $Co_xFe_{3-x}O_4$ (CF) were synthesized such as $Co_{0.2}Fe_{2.8}O_4$ (CF2), $Co_{0.4}Fe_{2.6}O_4$ (CF4), $Co_{0.5}Fe_{2.5}O_4$ (CF5), $Co_{0.6}Fe_{2.4}O_4$ (CF6), $Co_{0.8}Fe_{2.2}O_4$ (CF8), and $CoFe_2O_4$ (CF10). Similar procedure was employed for the synthesis of $Ni_xFe_{3-x}O_4$ materials.

Characterization of the derived ferrites

Phase purity, crystallite size, morphology, specific surface area, and elemental composition of the sol-gel derived ferrite powders were analyzed using a Panalytical XPert MPD/DY636 powder X-ray diffractometer, a Zeiss Supra 55VP field-emission scanning electron microscope (SEM), a FEI – Tecnai G2 200kV transmission electron microscope (TEM), a BET surface area analyzer (Micromeritics, ASAP 2420), and an ICP (Thermo, iCAP 6500).

Thermochemical CO_2-splitting set-up and procedure

Cycles consisting of the thermal reduction and the subsequent re-oxidation by CO_2 were performed with the sol-gel derived ferrites using a thermogravimetric analyzer (TGA, Netzsch STA 409) as shown in Figure 1b. Approximately 50 mg of ferrite powder was placed in an Al_2O_3 crucible. The thermal reduction was performed at 1400°C in a flow of 100 ml/min Ar. The samples were kept for 60 min at the reduction temperature. After thermal reduction step, the temperature was lowered to 1000°C to perform the CO_2 splitting reaction. The CO_2 was admitted (100 ml/min, Ar:CO_2=1:1) for 30 min to induce the re-oxidation of the ferrites. The mass loss/increase observed is correlated with the O_2 release/uptake of the sample during the reduction/re-oxidation step. O_2 and CO evolution was determined by on-line gas chromatography (GC, VARIAN, CP-4900, Micro GC 2 channel system)) of the off gas flow.

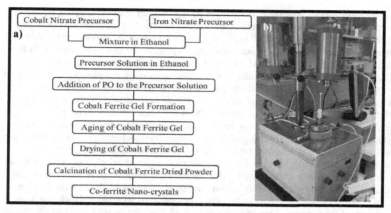

Figure 1. a) Sol-gel route exemplified for the synthesis of $Co_xFe_{3-x}O_4$, b) TGA Setup (Netsch STA409) used to assess the performance of ferrites during thermochemical CO_2-splitting cycles.

RESULTS & DISCUSSION

The phase purity of the derived $Ni_xFe_{3-x}O_4$ and $Co_xFe_{3-x}O_4$ materials was analyzed using powder x-ray diffractometer. For instance, the X-ray diffraction patterns of $Co_xFe_{3-x}O_4$ are shown in Figure 2a. In all patterns presented, strong reflections indicating a well-defined spinel cubic structure with high degree of crystallinity were found. No reflections that could be attributed to impurities such as precursors, CoO or metallic cobalt were observed. With the increase in the doping level of the cobalt metal into the ferrite spinel structure, a successive shift in the diffraction patterns of the $Co_xFe_{3-x}O_4$ materials towards lower angle was observed. Similar to the $Co_xFe_{3-x}O_4$ materials, $Ni_xFe_{3-x}O_4$ materials synthesized during this investigation also exhibit strong XRD reflections indicating highly crystalline spinel cubic structure with no evidence of any impurities such as NiO, FeO, or metallic Ni. The elemental composition of the synthesized ferrite materials were further confirmed with the help of ICP analysis.

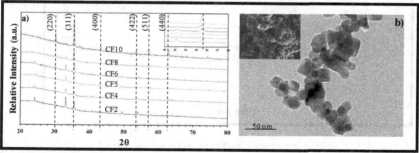

Figure 2. a) XRD patterns of sol-gel derived $Co_xFe_{3-x}O_4$ (where, x = 0.2 to 1), and b) SEM and TEM images of CF10.

The size of the single crystalline domains of sol-gel derived $Ni_xFe_{3-x}O_4$ and $Co_xFe_{3-x}O_4$ materials were calculated by using Scherrer equation. This quantitative analysis indicated slight decrease in the crystallite size of the $Ni_xFe_{3-x}O_4$ and $Co_xFe_{3-x}O_4$ with the increase in the degree of substitution of Ni and Co in the ferrite spinel structure. Typically, the crystallite sizes lie in the range 25 to 30 nm. Furthermore, the SEM and TEM analyses confirm the results of the XRD analysis. The materials were highly crystalline and exhibit nano-sized morphology with a size of the primary particles in the range 20 to 30 nm. As example, SEM and TEM pictures of CF10 are reported in Figure 2b. The specific surface area of the sol-gel derived $Ni_xFe_{3-x}O_4$ and $Co_xFe_{3-x}O_4$ obtained after calcination at 600°C was observed to be in the range of 35 to 40 m²/g.

The suitability of the derived ferrites for their use in thermochemical H_2O and CO_2 splitting cycles was assessed by performing four consecutive thermochemical cycles using a high temperature thermogravimetric analyzer (see Figure 1b). The experimental procedure is described in previous section. A baseline run was conducted under identical conditions as in the experiment but in absence of a ferrite sample. This allows subtracting artifacts due to buoyancy effects or caused by changing the gas composition. In Figure 3 we report data obtained during four cycles of CF10, CF8 and CF6 as example. The temperature is reported as black line while the weight change of the samples is reported in color. In all cases the thermal reduction seems to occur at a much slower rate than the re-oxidation. The thermal equilibrium composition is not reached even after 60 min. Re-oxidation is relatively fast and seems to be mostly completed with 30 min. Within the four cycles presented the redox capacity of the samples does not deteriorate i.e. approximately the same mass loss/increased is observed in all cycles. Note, that during the first reduction a disproportionally large weight loss occurs. We attribute a significant fraction of this weight loss to the desorption of physisorbed water from the samples.

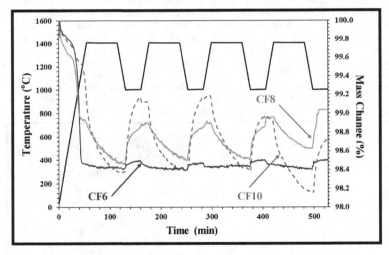

Figure 3. TGA of CO_2-splitting experiments performed using sol-gel derived $Co_xFe_{3-x}O_4$.

Amounts of O_2 released and CO produced in four consecutive thermochemical cycles performed using $Co_xFe_{3-x}O_4$ materials were calculated with the help of % weight changes (TGA) and the data obtained via gas chromatography and are reported in Table 1. The findings indicate that the O_2 released during the 1st thermal reduction step was much higher in case of CF2, CF4, and CF5 as compared to the other $Co_xFe_{3-x}O_4$ materials. However, in case of CF10, the O_2 released was observed to be the highest and constant during the 2nd, 3rd, and 4th thermochemical cycles. During the CO_2-splitting steps, similar trends were observed in terms of the CO produced by CF10 as compared to other CF materials. The average thermal reduction and CO_2-splitting ability of the derived $Co_xFe_{3-x}O_4$ materials during 2nd, 3rd, and 4th thermochemical cycles was observed to be in the order of CF10 > CF8 > CF6 > CF5 > CF4 > CF2, respectively. Currently, experiments are underway towards investigating the thermal reduction and CO_2-splitting ability of the sol-gel derived $Ni_xFe_{3-x}O_4$ materials during multiple thermochemical cycles.

Table 1. Amounts of O_2 released and CO produced by CF materials during thermal reduction and CO_2-splitting experiments performed in the temperature range of 1000 – 1400°C.

Material	Amounts of O_2 released/CO produced (µmol/g)							
	1st cycle		2nd cycle		3rd cycle		4th cycle	
	O_2	CO	O_2	CO	O_2	CO	O_2	CO
CF2	768	94	22	75	19	62	25	44
CF4	747	100	47	75	41	56	28	50
CF5	484	113	53	94	47	69	38	56
CF6	256	125	59	113	56	94	53	75
CF8	197	263	116	225	103	213	78	200
CF10	372	494	241	469	234	369	244	363

Previously, several ceria based redox materials were examined towards CO_2 splitting applications [19-23]. Among these materials $Ce_{0.75}Ca_{0.05}Zr_{0.2}O_{1.95}$ is capable of producing maximum amount of CO (259 µmol/g·cycle) and O_2 (146 µmol/g·cycle) in three consecutive thermochemical cycles performed in the temperature range of 1100 – 1400°C. The results obtained during this investigation indicate that the stoichiometric CF10 material performs best when compared with the other $Co_xFe_{3-x}O_4$ materials. An average of 273 µmol/g·cycle O_2 and 424 µmol/g·cycle of CO were produced with CO to O_2 ratio equal to 1.55. These values are higher than those reported for ceria based redox materials studied previously. Furthermore, the temperatures employed in this study are lower as compared to the previous investigations.

Sol-gel derived $Co_xFe_{3-x}O_4$ materials exhibit a superior redox capacity in terms of the O_2 released and CO produced when compared to other materials such as pure and doped ceria based materials which are under consideration as an active material in STCs for the production of solar fuels [23]. However, to make ferrite based STC economically viable, it is highly essential that these materials do not deactivate and produce high and constant levels of solar fuels in multiple STCs with superior kinetics as compared to the ceria based materials. Therefore, efforts are underway to study the reactivity, thermal stability, and reaction kinetics of the sol-gel derived ferrite materials on a much longer time scale (tens to hundreds of thermochemical cycles).

CONCLUSIONS

In this investigation, $Ni_xFe_{3-x}O_4$ and $Co_xFe_{3-x}O_4$ (where, x = 0.2 to 1) redox materials were successfully synthesized via sol-gel method by using nitrate salts of Ni, Co, and Fe, ethanol as solvent, and PO as gelation agent. The XRD analysis confirms the phase pure formation of ferrite materials with high degree of dopant incorporation in the ferrite spinel structure and with no impurities. Quantitative XRD and SEM/TEM analysis reveal nanocrystalline morphology of the derived ferrites. As-synthesized ferrite powders were further investigated for their use in thermochemical CO_2-splitting and thermal reduction cycles in the temperature range 1000 to 1400°C using a thermogravimetric analyzer. The results obtained indicate that the stoichiometric CF10 generates the largest amounts of O_2 and CO when compared to the other ferrites of this study and other non-volatile redox materials such as ceria based oxides investigated previously.

ACKNOWLEDGMENTS

The authors gratefully acknowledge the financial support provided by the Qatar Society of Petroleum Engineers (QSPE), the Indo-Swiss Joint Research Program (ISJRP, grant #138852), and the Swiss Federal Office of Energy (SFOE).

REFERENCES

1. Woods Hole Research Center, http://www.whrc.org/resources/primer.html (2014) last accessed April 1, 2014.
2. US Environment Protection Agency, http://epa.gov/climatechange/ (2014) last accessed April 1, 2014.
3. R. Bhosale, V. Mahajani, *Sep. Sci. Technol.* **48**, 2324 (2013).
4. R. Bhosale, V. Mahajani, *J. Renewable Sustainable Energy* **5**, 063110-1 (2013).
5. A. Gal, S. Abanades, G. Flamant, *Energy Fuels* **25**, 4836 (2011).
6. A. Steinfeld, *Sol. Energy* **78**, 603 (2005).
7. T. Kodama, N. Gokon, R. Yamamoto, *Sol. Energy* **82**, 73 (2008)
8. J. Scheffe, J. Li, A. Weimer, *Int. J. Hydrogen Energy* **35**, 3333 (2010).
9. M. Roeb, N. Gathmann, M. Neises, C. Sattler, R. Pitz-Paal, *Int. J. Hydrogen Energy* **33**, 893 (2009).
10. S. Lorentzou, C. Agrafiotis, A. Konstandopoulos, *Granular Matter* **10**, 113 (2008).
11. S. Stenger, M. Neises, M. Roeb, C. Sattler, *Energy Techol. TMS* (2011).
12. K. Allen, N. Auyeung, N. Rahmatian, J. Klausner, E. Coker, *JOM* **65**, 1682 (2013).
13. K. Allen, E. Coker, N. Auyeung, J. Klausner, *JOM* **65**, 1670 (2013).
14. R. Bhosale, R. Shende, J. Puszynski, *J. Energy Power Eng.* **4**, 27 (2010).
15. R. Bhosale, R. Shende, J. Puszynski, *Int. Rev. Chem. Eng.* **2**, 852 (2010).
16. R. Bhosale, R. Shende, J. Puszynski, *Int. J. Hydrogen Energy* **37**, 2924 (2012).
17. R. Bhosale, R. Khadka, R. Shende, J. Puszynski, *J. Renewable Sustainable Energy* **3**, 063104-1 (2011).
18. R. Bhosale, R. Shende, J. Puszynski, *Mater. Res. Soc. Symp. Proc.* 1387 (2012).
19. M. Kang, X. Wu, J. Zhang, N. Zhao, W. Wei, Y. Sun, *RSC Advances* **4**, 5583 (2014).
20. Q. Jiang, G. Zhou, Z. Jiang, C. Li, *Sol. Energy* **99**, 55 (2014).
21. M. Kang, J. Zhang, C. Wang, F. Wang, N. Zhao, F. Xiao, W. Wei, Y. Sun, *RSC Advances* **3**, 18878 (2013).
22. S. Abanades, A. L. Gal, *Fuel* **102**, 180 (2012).
23. P. Furler, J. Scheffe, A. Steinfeld, *Energy & Environmental Science* **5**, 6098 (2012).

Mater. Res. Soc. Symp. Proc. Vol. 1675 © 2014 Materials Research Society
DOI: 10.1557/opl.2014.867

Synthesis of Imprinted Polysiloxanes for Immobilization of Metal ions

Adnan Mujahid[1*], Faisal Amin[1], Tajamal Hussain[1], Naseer Iqbal[2,3], Asma Tufail Shah[2], Adeel Afzal[2,4]

[1] Institute of Chemistry, University of the Punjab, Quaid-i-Azam Campus Lahore-54590, Pakistan

[2] Interdisciplinary Research Centre in Biomedical Materials, COMSATS Institute of Information Technology, Defence Road, Off. Raiwind Road, Lahore 54000, Pakistan

[3] Department of Biosciences, COMSATS Institute of Information Technology, Park Road ,Chak Shahzad,Islamabad 45600, Pakistan

[4] Affiliated Colleges at Hafr Al-Batin, King Fahd University of Petroleum and Minerals, P.O. Box 1803, 31991 Hafr Al-Batin, Saudi Arabia

* Email address: adnanmujahid.chem@pu.edu.pk Tel.: +92-42-99230463

ABSTRACT

Imprinting is a well-established technique to induce recognition features in both organic and inorganic materials for a variety of target analytes. In this study, ion imprinted polysiloxanes with varying percentage of coupling agent i.e. 3-chloro propyl trimethoxy silane (CPTM) were synthesized by sol-gel method for imprinting of Cr^{3+}. The imprinting of Cr^{3+} in cross-linked siloxane network was investigated by FT-IR which indicates the metal ion is coordinated with oxygen atoms of polysiloxanes. SEM images revealed that imprinted polysiloxanes possess uniform particles of submicron size. It was experienced that by increasing the concentration of CPTM up to 10% (v/v) substantially improves the binding capacity of polysiloxanes which allows us to recognized Cr^{3+} down to 50µg/L. Furthermore, the selectivity of Cr^{3+}-imprinted polysiloxanes was evaluated by treating them with other competing metal ions of same concentration i.e. Cr^{6+}, Pb^{2+} and Ni^{2+}. In this regard, polysiloxanes showed much higher binding for imprint ion i.e. Cr^{3+} in comparison to above mentioned metal ions. Finally, the regenerated polysiloxanes were studied in order to reuse them thus, developing cost effective biomimetic sensor coatings.

INTRODUCTION

During the last few decades, the increasing use of different heavy metals pose serious environmental and health hazards. Metal smelting, tanning, electroplating, metallurgy for steel production and paint industries are major contributors of heavy metal ions in environment. Apart from others, chromium [1] is an important element in making of alloy steel, electroplating and many other industrial processes. The waste from these industries contains high concentration of chromium and usually discharged without proper disposal. This results in contamination of surface and ground water as well as drinking water [2].

Generally, there are two main approaches for chromium removal from waste water i.e. sorption [3-5] of target metal ions onto various materials and other is separation through membrane filtration [6, 7]. Nevertheless, sorption is relatively more convenient and direct method for removing toxic materials as here the metal ions are adsorbed on the surface of solid absorbent. The amount of metal ions removed depends upon different parameters, mainly adsorption capacity of sorbent, pH, temperature and others. The most important aspect is the selectivity of the sorbent for target metal ion. This concern leads to the development of new synthetic materials for ion selective separation applications. Recently, ion-imprinted polymers [8, 9] have shown considerable potential for selective separation and pre-concentration of toxic metal ions. These materials have been used both as synthetic sorbent as well as for membrane separations. Enhanced sensitivity and selectivity, straightforward synthesis, high chemical stability and easy regeneration make them greatly desirable in various applications [10-12].

In this study, ion-imprinted polysiloxanes were synthesized by sol-gel method for selective recognition of Cr^{3+} in aqueous samples. Tetraethyl orthosilicate (TEOS) was taken as functional monomer whereas CPTM was attached during hydrolysis in different proportions (v/v) i.e. 5% and 10%. The imprinting of Cr^{3+} in polysiloxanes was accessed by FTIR studies and surface morphology was investigated by SEM. In rebinding studies, imprinted polysiloxanes exhibited high sensitivity and selectivity. The non-imprinted material exhibited negligible binding. Moreover, the reusability of imprinted polysiloxanes was also studied for their repeated use in binding experiments.

EXPERIMENTAL SECTION

All the chemicals and reagents were purchased from Sigma Aldrich and used as received without any prior treatment.

A mixture of TEOS and CPTM having composition of 95% (v/v) and 5% (v/v) respectively was stirred together in ethanol at 60°C. For imprinting Cr^{3+}, 5 mg of chromium nitrate dissolved in 5mL of deionized water was added into above matrix. Liquid ammonia was added drop wise to initiate and propagate hydrolysis reaction as it also leads to condensation of polysiloxanes. After gel formation, the reaction was allowed to continue for 2h under constant stirring. The gel was separated by centrifugation and dried in oven at 50°C for 4-5 h. The dried gel was subjected to washing for complete removal of Cr^{3+}. Second batch of polysiloxanes were prepared in the exactly same manner keeping the composition of TEOS and CPTM 90% (v/v) and 10% (v/v), respectively. Non-imprinted polysiloxanes were prepared following the same procedure except adding template ion.

Imprinting of Cr^{3+} in polysiloxanes was investigated by FTIR analysis for observing polymer template interactions. The spectrum was recorded on Perkin Elmer Spectrum BX. Surface morphology of ion-imprinted polysiloxanes was studied on a scanning electron microscope Hitachi S–4700. In rebinding experiments, the concentration of chromium solution before and after treating with imprinted polysiloxanes was determined by ICP-OES Perkin Elmer Optima 2100.

RESULTS AND DISCUSSION

Ion-imprinted polysiloxanes were first characterized by FTIR before and after washing by deionized water to remove template. It was observed that prior to washing there was a peak near 698 cm⁻¹ which suggests that Cr^{3+} is coordinated by oxygen moieties of polysiloxanes network [13,14] and thus indicated successful imprinting. After through washing of polymers, the metal-oxygen peak was missing in spectrum indicating the removal of template ions. The rest of structural features in polysiloxanes were similar before and after template washing. FTIR spectrum of imprinted polysiloxanes before and after template washing is shown in figure 1.

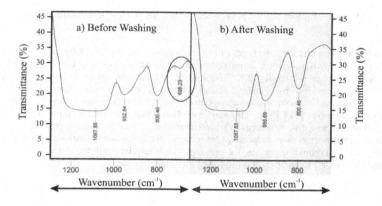

Figure 1: FTIR spectrum of imprinted polysiloxanes (a) prior to washing, (b) after template removal.

Surface morphology of imprinted polysiloxanes was studied by SEM image as shown in figure 2. This image shows that ion-imprinted polysiloxanes possess sub-micron particles which are distributed uniformly in the structure. SEM micrograph suggests that such surface is favorable for adsorption of target metal ion.

The imprinted polysiloxanes were subjected to several washing steps as to ensure the complete removal of template ions. It was observed that polysiloxanes containing 10% CPTM required relatively more time for template washing as compare to polysiloxanes comprises of 5% CPTM. After each washing step, the left over Cr^{3+} in aqueous medium was determined by ICP-OES. In rebinding experiments, it was observed that imprinted polysiloxanes containing 10% of CPTM showed enhanced affinity than 5% CPTM for Cr^{3+} recognition.

Figure 2: SEM image of ion-imprinted polysiloxanes showing the uniform distribution of particles.

The binding responses of two types of materials were compared at 50,100 and 200 µg/L of Cr^{3+} solution as shown in figure 3. Imprinted polysiloxanes having 10% of CPTM showed obvious change in emission intensity of Cr^{3+} solution at 50 µg/L whereas polysiloxanes having 5% of CPTM did not exhibit any change in emission intensity at this concentration of Cr^{3+}. Even at 100 and 200 µg/L of Cr^{3+} solution, the change in emission intensity is much higher for 10% CPTM polysiloxanes than 5% CPTM. This can be explained that increased proportion of CPTM lead to develop more sensitive structures in which number of interaction sites is greater. This feature make such materials more responsive for recognizing lower concentrations of target metal ion e.g. Cr^{3+}. In rebinding studies, the non-imprinted material exhibited much less response in binding Cr^{3+} as compared to imprinted material.

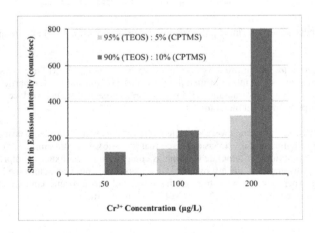

Figure 3: Rebinding results of two different ion-imprinted polysiloxanes at different concentrations of Cr^{3+}.

From these results, it is evident that polysiloxanes with increased percentage of CPTM are more efficient in rebinding experiments. Therefore, in following experiment we preferred this composition for selectivity studies where Cr^{3+}-imprinted polysiloxanes was added to different metal ions solution i.e. Cr^{6+}, Ni^{2+} and Pb^{2+} of equal concentrations and volume. The change in emission intensity of these metal ion solution was recorded before and after treating with imprinted polysiloxanes. The relative shift in emission intensity for these metal ions solution has been shown in figure 4.

This result clearly indicates that imprinted polysiloxanes are highly selective towards templated ion i.e.Cr^{3+} as compare to other competing ions. This feature makes ion-imprinted polysiloxanes favorable for selective separation of target metal ions comparing to other materials. Finally, we also tested the reusability of imprinted polysiloxanes. For this purpose, polysiloxanes used in binding experiments were regenerated and used in subsequent binding studies. It was noticed that the binding efficiency of regenerated polysiloxanes was more than 90%, which means that this material can be recycled and reused for several analyses without losing recognition performance.

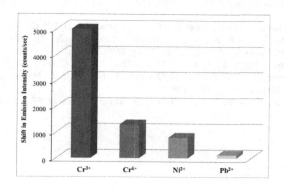

Figure 4: Relative shift in emission intensity of different metal ions solution i.e. Cr^{3+}, Cr^{6+}, Ni^{2+} and Pb^{2+} when treated with ion-imprinted polysiloxanes, highest shift in intensity was observed for template ion i.e. Cr^{3+}.

CONCLUSION

In the present work, we experienced that ion-imprinted polysiloxanes are highly suitable sorbent materials for separation of toxic metal ions from water samples. Their enhanced sensitivity and selectivity make them vastly useful as compare to other sorbents. The optimal composition of polymer system has important role in improving binding efficiency of ion-imprinted sorbent. Their easy regeneration is useful for several round of rebinding. Not only in separation and membrane filtration but they can also be used as efficient sensor coating material for recognition of target metal ions.

REFERENCES

1. J. Kotas, Z. Stasicka, Environmental Pollution, 107 (2000) 263-283.
2. D. Metze, N. Jakubowski, D. Klockow, Speciation of Chromium, in: Handbook of Elemental Speciation II – Species in the Environment, Food, Medicine and Occupational Health, John Wiley & Sons, Ltd, 2005, pp. 120-135.
3. N. K. Lazaridis, K. A. Matis, M.Webb, Chemosphere 42 (4), 373 (2001).
4. N. K. Hamadi, X. D.Chen, M. M. Farid, M. G. Q. Lu, Chemical Engineering Journal 84 (2), 95 (2001).
5. S. Rengaraj, K.-H. Yeon, S.-H. Moon, Journal of Hazardous Materials 87 (1–3), 273 (2001).
6. A. Aliane, N. Bounatiro, A.T. Cherif, D.E. Akretche, Water Research 35 (9), 2320 (2001).
7. A. Bhowal,S. Datta, Journal of Membrane Science 188 (1), 1 (2001).
8. T. Prasada Rao, s. Daniel, J. Mary Gladis, TrAC Trends in Analytical Chemistry 23 (1), 28 (2004).
9. T. P. Rao, R. Kala, S. Daniel, Analytica Chimica Acta 578 (2), 105 (2006).
10. L. Mafu, T.M. Msagati, B. Mamba, Environ. Sci. Pollut. Res. 20 (2), 790 (2013).
11. C. Branger, W. Meouche, A. Margaillan, 73(6), 859 (2013).
12. H. He, D. Xiao, J. He, H. Li, H. He, H. Dai, J. Peng, Analyst (2014).
13. J. S. T. Mambrim, H. O. Pastore, C. U. Davanzo, E. J. S. Vichi, O. Nakamura, H. Vargas, Chemistry of Materials 5(2), 166 (1993).
14. H. Fuks, S. Kaczmarek, M. Bosacka, Rev. Adv. Mater. Sci, 23(1), 57 (2010).

Mater. Res. Soc. Symp. Proc. Vol. 1675 © 2014 Materials Research Society
DOI: 10.1557/opl.2014.889

Chemical Solution Based MoS$_2$ Thin Film Deposition Based on Dimensional Reduction

Changqing Pan[1,2] , Zhongwei Gao[1] and Chih-hung Chang[1,2]
[1]School of Chemical, Biological, and Environmental Engineering, Oregon State University, Corvallis, OR 97331, U.S.A.
[2]Oregon Process Innovation Center, Corvallis, OR 97330, U.S.A.

ABSTRACT

As a promising transition metal dichalcogenide (TMDC), molybdenum disulfide (MoS$_2$) has recently attracted a lot of attention due to its graphene-liked two dimensional layer structure, which leads to potential applications in electronic and optoelectronic devices. However, the fabrication of mono- or few-layer MoS$_2$ is limited to ether liquid exfoliation or CVD, and the chemical solution deposition is limited to ammonium thiomolybdate-based precursor. In this paper, hydrazine-based dimensional reduction technique is applied in the chemical solution deposition of MoS$_2$ thin-film, and a larger area uniform thin-film is obtained from bulk powder MoS$_2$. This solution-based process could be applied with a variety coating techniques and lead to wafer level MoS$_2$ thin film production.

INTRODUCTION

Graphene has attracted wide attention due to its extreme high carrier mobility and other properties connected with the 2-D structure.[1,2] As the graphene analogues, mono-layer or few-layer transition metal dichalcogenides (TMDCs) such as MoS$_2$, WS$_2$, and SnS$_2$ also gain the new attention after decades of study on the bulk materials. Among these TMDCs, MoS$_2$ is one of the most promising candidates for microelectronics applications due to its high field-effect mobility and high on/off ratio.[3]

In order to actually integrate the few-layer MoS$_2$ thin-film in field effect thin film transistor, the reliable production of larger area uniform MoS$_2$ thin films becomes a critical issue to solve. Current synthesis methods of MoS$_2$ thin film include chemical exfoliation, chemical vapor deposition (CVD) and chemical solution deposition (CSD).[4] Chemical exfoliation is a quick way to obtain MoS$_2$ nanoflakes or nanosheets from bulk molybdenite crystals, but the size of these flakes varies and there is also a challenge to align these flakes to desired position for devices fabrication.[5] CVD has been used to synthesis MoS$_2$ thin-film on reduced graphene oxide (rGO) treated substrate from MoO$_3$ and elemental sulfur precursors, with a relatively low material utilization.[6] Chemical solution deposition shows a high material utilization ratio and could achieve a large area uniform MoS$_2$ thin film, but currently the chemicals are limited to ammonium thiomolybdate and alkyldiammonium thiomolybdate.[7]

Dimensional reduction is an alternative approach for metal chalcogenide thin film deposition.[8] By using hydrazine (N$_2$H$_4$) as dimensional reduction agent, ultra-thin SnS$_2$ film for TFT[9] and champion efficiency CuZnSnSSe solar cell absorber layer were obtained.[10] MoS$_2$ as a metal dichalcogenide can also be dissolved in N$_2$H$_4$ with sulfur, which opens a door for transformation of MoS$_2$ bulk powder material into few-layer thin film structure. Since this process involves N$_2$H$_4$ which is highly toxic and explosive, a micro-reactor made from stainless steel and polyvinylidene fluoride (PVDF) is used to maximum limit the exposure of N$_2$H$_4$.

EXPERIMENT

The experiment setup is shown in Figure 1. 0.320 g MoS$_2$ bulk powder (≤ 2 µm, Aldrich) and 0.160 g sulfur powder were mixed and loaded into a micro-reactor chamber, N$_2$H$_4$ (anhydrate, Aldrich) was injected into the microreactor by syringe (Figure 1.a). The product solution from outlet of the micro-reactor was collected and used for spin-coating on glass substrate (Figure 1.b). The film then was heated at 200 °C to remove the exceed N$_2$H$_4$ solvent and transform to MoS$_3$ film (Figure 1.c). After a second annealing at 400 °C with 10% H$_2$ in N$_2$ for 40 minutes (with extra ramping time at 20 °C/min), the film was converted into final product: MoS$_2$ (Figure 1.d).

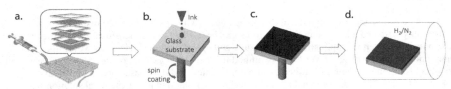

Figure 1. The Schematic Diagram of Experiment Setup.

The proposed reaction for dimensional reduction is shown in Table 1:

Step a.	$2S + 5N_2H_4 + MoS_2 \rightarrow MoS_4^{2-} + 4[N_2H_5]^+ + N_2 \uparrow$
Step b.	$[N_2H_5]_2 MoS_4 \xrightarrow{200°C} MoS_3 + 2N_2H_4 \uparrow + H_2S \uparrow$
Step c.	$MoS_3 + H_2 \xrightarrow{400°C} MoS_2 + H_2S$

Table 1. The chemical reactions during the dimensional reduction process.

DISCUSSION

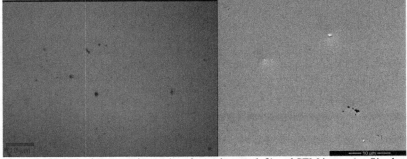

Figure 2. Optical microscopy image (on glass substrate, left) and SEM image (on Si substrate, right) of MoS$_2$ thin films.

Both optical microscope and Scanning Electron Microscope (SEM) were used to check the uniformity of the MoS₂ thin film, as shown in Figure 2. Most area of the substrate are uniformly covered with MoS₂, with only very few defects were found on the film in a relatively large film area,

Raman spectrum is one of the most powerful tools for the characterization of few-layer MoS₂ thin-film. The frequencies of E_{2g}^1 and A_{1g} peaks can be used as key features to determine the layer number of MoS₂ thin-film.[11] Raman spectra of both single layer spin-coated MoS₂ thin film and twice spin-coated film were measured through Witec Alpha Raman spectrometer. According to the peak positions, both films were multi-layer MoS₂.

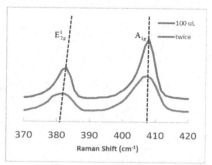

Figure 3. Raman shift of E_{2g}^1 and A_{1g} peaks from two MoS₂ thin film: singer layer spin coating (blue) and spin coating for twice (orange).

CONCLUSIONS

Bulk MoS₂ powder was successfully transformed into large area uniform MoS₂ thin films through dimensional reduction process with N₂H₄ as dimensional reduction agent. It could be used for further microelectronics devices fabrication and lead to a potential high mobility and on/off ratio transistor. Future study on the reaction mechanisms will include the TGA/DSC study on the precursors and intermediate products, the topography study of the film surface by atomic force microscope.

ACKNOWLEDGMENTS

The SEM images were obtained under the support of the OSU Electron Microscopy Facility. The authors also thank the support of Oregon BEST through Oregon Process Innovation Center.

REFERENCES

1. A.K. Geim, Science **324**, 1530 (2009).
2. A.S. Mayorov, et al., Nano Lett. **11**, 2396 (2011).
3. B. Radisavljevic, et al., Nat.Nanotechnol. **6**, 147 (2011).
4. Q.H. Wang, et al., Nat. Nanotechnol. **7**, 699 (2012).
5. J.N. Coleman et al., Science **331**, 568 (2011).
6. Y.-H. Lee, et al., Adv. Mater. **24**, 2320 (2012).
7. K.-K. Liu, et al., Nano Lett. **12**, 1538 (2012).
8. D.B. Mitzi, Adv. Mater **21**, 3141 (2009).
9. D.B. Mitzi, et al., Nature **428**, 299 (2004).
10. W. Wang, et al., Adv. Energy Mater. doi: 10.1002/aenm.201301465
11. H. Li, et al., Adv. Funct. Mater. **22**, 1385 (2012).

Mater. Res. Soc. Symp. Proc. Vol. 1675 © 2014 Materials Research Society
DOI: 10.1557/opl.2014.888

Organometallic synthesis of water-soluble ruthenium nanoparticles in the presence of sulfonated diphosphines and cyclodextrins

Miguel Guerrero,[1] Nguyet Trang Thanh Chau,[2] Alain Roucoux,[2] Audrey Nowicki-Denicourt,[2] Eric Monflier,[3] Hervé Bricout[3] and Karine Philippot*[1]

[1] CNRS; LCC (Laboratoire de Chimie de Coordination); 205 Route de Narbonne, F-31077 Toulouse, France; Université de Toulouse; UPS, LCC; F-31077 Toulouse; France.
[2] Ecole Nationale Supérieure de Chimie de Rennes, UMR, CNRS 6226, Avenue du General Leclerc, CS 50837, 35 708 RENNES Cedex 7; Université Européenne de Bretagne; France.
[3] UMR CNRS 8181; Université d'Artois, Rue Jean Souvraz, F-62307 Lens Cedex; France.

*Corresponding author; E-mail address: karine.philippot@lcc-toulouse.fr

ABSTRACT

The organometallic approach was successfully applied to synthesize water-soluble ruthenium nanoparticles displaying interesting catalytic properties in hydrogenation of unsaturated model-substrates. Nanocatalyst synthesis was performed by hydrogenation of the complex [Ru(COD)(COT)] in the presence of sulfonated diphosphines and cyclodextrins as protective agents providing very small ruthenium nanoparticles (ca. 1.2-1.5 nm) with narrow size distribution and high stability. Catalysis results in water evidenced a control of the surface properties of these novel ruthenium nanocatalysts at a supramolecular level.

INTRODUCTION

Since the end of the 1990s, and with the development of nanosciences, nanocatalysis has clearly emerged as a domain at the interface between homogeneous and heterogeneous catalysis, which offers unique solutions to answer the demanding conditions for catalyst improvement.[1,2] A modern approach of colloidal chemistry is presently being developed to increase the reactivity of nanoparticles in a limited size up to 10 nm, using several types of capping agents.[3] Among others the organometallic approach represents an efficient synthetic methodology to get well-controlled metallic nanoparticles in terms of size and composition, two key-parameters for application in nanocatalysis.[4] Various metal organic complexes can be used as metal source, being decomposed under mild conditions of solution chemistry, to provide mono- and bimetallic systems. This approach is also versatile in terms of stabilizing agents since polymers, ligands, ionic liquids or inorganic supports (alumina, silica, carbone derivatives, etc) can be used. The choice of the stabilizer is critical as it will govern nanoparticle characteristics as their size, shape, dispersion, surface properties and further, influence their catalytic properties. In this context, choosing ligands as those used in organometallic chemistry and catalysis is thus of interest since one can expect to tune the surface properties of metal nanoparticles as performed with molecular complexes. This can be even applied to obtain water-soluble nanoparticles while starting their synthesis in organic medium as firstly evidenced using 1,3,5-triaza-7-

phosphaadamantane (PTA) for the preparation of platinum and ruthenium nanoparticles in water.[5,6] Moreover, biphasic liquid-liquid systems are an attractive alternative for recovering aqueous soluble nanocatalysts after separation from the organic phase containing products, as previously done for homogeneous catalysts.[7,8]

To develop new aqueous nanocatalytic systems, we chose sulfonated diphosphines as capping ligands for the synthesis of Ru(0) nanoparticles (RuNPs). The aim was to take advantage of the strong coordination of sulfonated ligands at the metal surface and of their high water-solubility to have stable RuNPs and to transfer them easily in water. The nanoparticles were then prepared in THF in the presence of sulfonated diphosphines leading first to stable organic suspensions. The RuNPs were isolated as powders and easily dispersed by simple addition of water, leading to aqueous colloidal solutions containing RuNPs with low size dispersity and very small mean diameters (ca. 1.2-1.5 nm) that were stable for several months. These systems were involved in the hydrogenation of model unsaturated substrates (arenes and alkenes) in biphasic aqueous-organic conditions.[9] The results showed that the reaction selectivity could be tuned in favor of target products depending on the reaction conditions. To increase and modulate the catalytic performances of our Ru nanocatalysts we also considered the addition of a suitable molecular receptor namely a cyclodextrin. This was inspired by previous works where cyclodextrins were successfully used as mass-transfer promoters to improve a catalytic reaction in aqueous/organic biphasic conditions using organometallic complexes or ruthenium nanoparticles prepared by reduction of ruthenium trichloride salt respectively. [10,11] Spectroscopic studies have demonstrated that β-CD and RAME-β-CD (randomly methylated-beta-*cyclodextrin*) interact with sulfonated phosphines by forming inclusion complexes, thus tuning the catalytic performances of the metal centers.[12] Thus, different combinations between a sulfonated diphosphine (1,4-bis [(di-m-sulfonatophenyl) phosphino]butane) and RAME-β-cyclodextrin added as co-additive were envisaged to prepare RuNPs. Our strategy relied on combining advantages of both a sulfonated diphosphine as efficient stabilizer for metal nanoparticles in aqueous solution and a cyclodextrin for its shuttle and supramolecular control effects in catalysis. With highly stable RuNPs in water it was possible to characterize for the first time the ligand/cyclodextrin association and to study its influence on the reactivity during hydrogenation reaction of aromatic model substrates, in terms of conversion/selectivity and of stability/recyclability.[13] Even if the synthesis method was different, the idea was to compare these new RuNPs with other RuNPs which were stabilized only with a cyclodextrin. [11]

EXPERIMENTAL DETAILS

General: Nanoparticle syntheses were carried out in Schlenck or Fischer-Porter glassware or in a glove-box under argon atmosphere. The organometallic complex used as precursor, (1,5-cyclooctadiene)(1,3,5-cyclooctatriene)ruthenium (0) complex ([Ru(COD)(COT)]; COD=1,5-cyclooctadiene; COT=1,3,5-cyclooctatriene) was purchased from Nanomeps-Toulouse. Sulfonated diphosphines (1,4-bis (1,4-bis [(di-m-sulfonatophenyl) phosphino]butane=L1, 1,4-bis [(di-m-sulfonatophenyl)phosphino]propane=L2 and 1,4-bis [(di-m-sulfonatophenyl)phosphino]ethane=L3) were synthesized following published procedure [14] and RAME-β-CD (randomly methylated-beta-*cyclodextrin*; Cavasol® W7 M) was purchased from Wacker Chemie GmbH in its pharmaceutical grade and was used as received. RAME-β-CD is a partially methylated β-cyclodextrin and its degree of substitution is equal to 1.8 per glucopyranose unit.[15]

Solvents were dried and distilled before use: tetrahydrofurane (THF) over sodium-benzophenone and pentane over calcium hydride. All reagents and solvents were degassed before use by means of three freeze-pump-thaw cycles. Water was distilled twice by conventional method to prepare nanoparticle suspensions. Styrene, acetophenone and methyl were purchased from Acros Organics or Sigma-Alfa Aesar and used as substrates in catalysis without further purification.

Synthesis of RuNPs: The ruthenium nanoparticles were prepared by hydrogenation of the organometallic precursor [Ru(COD)(COT)] in THF at room temperature in the presence of alkyl sulfonated diphosphine ligands (L) in different [ligand]/[metal] molar ratio (0.1, 0.2 or 0.5). After precipitation by addition of pentane and isolation, the obtained nanoparticles were easily transferred into water giving rise to stable aqueous colloidal solutions that were investigated in hydrogenation of unsaturated model substrates. The synthesis of RuNPs was also performed in the presence of both L1 and RAME-β-CD to study the influence of this transfer agent on the catalysis performances.

Characterization: The RuNPs were characterized in solution by liquid Nuclear magnetic resonance (NMR) and dynamic light scattering (DLS), after isolation of the particles and dispersion in water. Grids were prepared from THF and aqueous solutions for transmission electron microscopy (TEM) and high resolution TEM analysis. The purified nanoparticles as powders were also characterized by solid state NMR spectroscopy, elemental analysis, wide-angle X-ray scattering (WAXS).

RESULTS AND DISCUSSION

Contrarily to previous work where RuNPs were prepared by reduction of ruthenium trichloride as metal source in water [11], the synthesis of ruthenium nanoparticles was here performed by the hydrogenation of the organometallic complex [Ru(cod)(cot)] in THF under mild conditions (3 bar H_2; room temperature) in the presence of the chosen sulfonated diphosphine (1,4-bis [(di-m-sulfonatophenyl) phosphino]butane=L1, 1,4-bis [(di-m-sulfonatophenyl)phosphino]propane=L2 or 1,4-bis [(di-m sulfonatophenyl)phosphino]ethane=L3) or a combination of L1 with the RAME-β-cyclodextrin (CD) (Figure 1). The advantage of this method compared to the reduction of a metal salt [11] is the control of the metallic surface state which is assumed to be cleaner given that no pollution is expected. Indeed, the decomposition of the metal precursor leads to the liberation of alkanes which are inert towards the metal surface. In these conditions it is possible to finely study the surface reactivity by NMR studies, like the coordination of ligands for example.

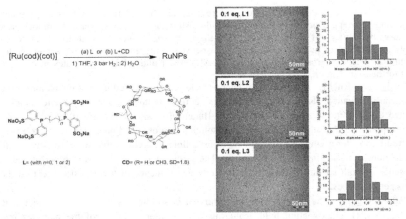

Figure 1. Synthesis of the RuNPs and TEM images from the colloidal aqueous phases

Whatever the sulfonated diphosphine used, very small and well-controlled RuNPs were formed in organic phase. Precipitation by addition of pentane allowed to isolate the RuNPs that were further easily dispersed in water taking profit of the solubility of the sulfonated diphosphine ligands in this solvent. The so-obtained aqueous colloidal solutions contained RuNPs of low size dispersity and mean diameters in the range 1.2-1.5 nm depending on the [L]/[Ru] ratio. No change in size was observed after transfer of RuNPs in water. NMR studies (in the solid state and on aqueous colloidal solutions) showed the strong coordination of sulfonated diphosphine ligands with RuNPs surface. This induced an efficient control of their size as well as of their surface properties and facilitated their transfer into water. In all cases, the ruthenium aqueous colloidal suspensions were found to be stable for several months. They were tested in the hydrogenation of model unsaturated substrates (arenes and alkenes) in biphasic aqueous-organic conditions and compared to RuNPs prepared via the salt reduction method [11]. The results showed active nanocatalysts despite the change of environment they underwent after their dispersion in water. It was also observed that the reaction selectivity could be tuned in favor of target products depending on the reaction conditions (temperature, H_2 pressure, [L]/[Ru] ratio, etc) with higher reactivity/selectivity than RuNPs synthesized from $RuCl_3$ in some cases. [11]

When the synthesis of RuNPs was performed in the presence of 1,4-bis [(di-m-sulfonatophenyl) phosphino]butane (L1) and RAME-β-cyclodextrin (CD), stable aqueous colloidal solutions containing very small RuNPs (ca. 1.5 nm) were also achieved. Given the ability of CD to form supramolecular inclusion entities with phosphine ligands of molecular complexes in aqueous media, our goal was to study if the CD could have a similar behavior with RuNPs and induce a modification of the diphosphine coordination at the metal surface. Thus, several quantities of RAME-β-CD (0.2, 1.0 and 5.0 equiv.) were investigated to evaluate the effect of CD both on the stability of the sulfonated diphosphine-stabilized RuNPs and on their catalytic properties during hydrogenation reaction of aromatic model substrates, in terms of conversion and selectivity. The existence of an interaction between L and CD was clearly evidenced by NMR spectroscopy (both in solution and in solid state) which appeared even stronger at higher CD content, thus disrupting severely the coordination properties of the ligand towards the metal surface. But we observed that this interaction took place only if the cyclodextrin was present during the synthesis of the nanoparticles. The aqueous nanocatalytic systems were investigated in the hydrogenation of styrene, and acetophenone and 1-methoxy-3-methylbenzene (*m*-methylanisole). Increasing TOF values were observed with increasing CD contents showing that CD-more rich nanocatalysts were more active, thus evidencing the shuttle effect of the cyclodextrin. Nevertheless, some differences were observed depending on the quantity of CD: at low quantity of CD (0.2 equiv.), the effect was low (probably due to the fact that CD was mainly involved in the formation of an inclusion complex with the diphosphine ligand at the metal surface, thus leading to stable RuNPs that are active but with no boosting effect); at high quantity of CD, there was enough free CD in the reaction medium to improve the catalytic system. Concerning the selectivity, a clear influence of the CD was visible in the hydrogenation of *m*-methylanisole (Figure 2). With this disubstituted aromatic substrate we could determine the influence of the CD on the reaction stereoselectivity as two diastereomeric products (cis/trans) can be formed. Besides increasing TOF values when high quantity of CD was present in the reaction medium, diastereomeric excesses between 50 and 100% were also noticed, in favor of the kinetic *cis* product. These results evidenced clearly that the quantity of CD influenced the selectivity of the reaction, the more CD rich nanocatalysts (Ru/L/5.0CD) being the more selective. Finally, encouraging results were obtained in the recovery/reusability of the aqueous nanocatalysts with similar reactivity.

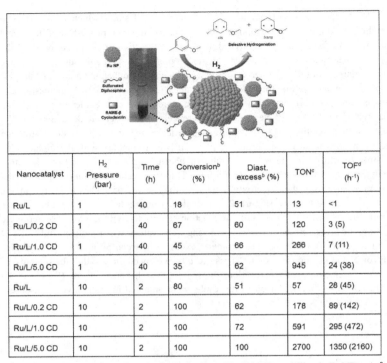

Nanocatalyst	H_2 Pressure (bar)	Time (h)	Conversion[b] (%)	Diast. excess[b] (%)	TON[c]	TOF[d] (h⁻¹)
Ru/L	1	40	18	51	13	<1
Ru/L/0.2 CD	1	40	67	60	120	3 (5)
Ru/L/1.0 CD	1	40	45	66	266	7 (11)
Ru/L/5.0 CD	1	40	35	62	945	24 (38)
Ru/L	10	2	80	51	57	28 (45)
Ru/L/0.2 CD	10	2	100	62	178	89 (142)
Ru/L/1.0 CD	10	2	100	72	591	295 (472)
Ru/L/5.0 CD	10	2	100	100	2700	1350 (2160)

<u>Figure 2</u>. Hydrogenation of *m*-methylanisole with L1/CD-stabilized RuNPs ([a] reaction conditions: Ruthenium NPs (10 mg), methylanisole (3.9×10^{-3} mol), 1 or 10 bar H_2, room temperature, water (10 mL); [b] Determined by GC analysis; [c] Initial turnover number (TON) is expressed as moles of the substrate converted per moles of metal; [d] Initial turnover frequency (TOF) is expressed as moles of the substrate transformed per moles of metal per h and in brackets TOF is corrected per moles of surface Ru atoms.

CONCLUSIONS

Highly stable aqueous colloidal suspensions containing well-controlled 1.2-1.5 nm RuNPs were achieved through an organometallic approach by using sulfonated diphosphines as capping ligands or a combination of a sulfonated disphosphine and RAME-β cyclodextrin. NMR studies highlighted the strong interaction of the sulfonated diphosphines with the RuNPs surface as well as the formation of supramolecular inclusion complexes between sulfonated diphosphines and cyclodextrins. These systems displayed pertinent catalytic performances in the hydrogenation of unsaturated substrates in biphasic aqueous-organic conditions. Small structural differences in the backbone of the diphosphine ligands were shown to influence the catalytic properties of the nanocatalysts. Moreover, relevant differences were observed in the presence of cyclodextrins. The CD acted as a phase-transfer promotor by increasing the activity of the reaction but also affecting the selectivity as the result of a supramolecular interaction (the formation of strong inclusion complexes between the ligand and/or the substrate within the cage of the cyclodextrin) at the metallic surface of the nanocatalysts. In comparison with RuNPs prepared by reduction of a metal salt, these systems offer a good control of the metallic surface state thanks to the organometallic synthesis method . This work may thus open new opportunities in the field of nanocatalysis.

ACKNOWLEDGMENTS

The authors thank V. Collière and L. Datas (LCC-CNRS and UPS-TEMSCAN) for TEM/HRTEM facilities, P. Lecante (CEMES-CNRS) for WAXS measurements and Y. Coppel and C. Bijani (LCC-CNRS) for NMR studies. This work was supported by CNRS and the Agence Nationale de la Recherche (ANR-09-BLAN-0194) for the financial support of the SUPRANANO program.

REFERENCES

1. J.M. Thomas, *ChemCatChem*, 2010, **2**, 127-132.
2. D. Astruc, Nanoparticles and Catalysis, (Ed). Wiley-VCH, 2008.
3. K. Philippot and P. Serp, Nanomaterials in Catalysis (Eds.), Wiley-VCH, Weinheim, 2013, 1-54.
4. C. Amiens, B. Chaudret, D. Ciuculescu-Pradines, V. Collière, K. Fajerwerg, P. Fau, M. Kahn, A. Maisonnat, K. Soulantica, K.Philippot, *New J. Chem.*, 2013, **37** (11), 3374 - 3401.
5. P.-J. Debouttière, V. Martinez, K. Philippot, B. Chaudret, *Dalton Trans.*, 2009, 10172-10174.
6. P.-J. Debouttière, Y. Coppel, A. Denicourt-Nowicki, A. Roucoux, B. Chaudret, K. Philippot, *EurJIC.*, 2012, 1229-1236.
7. B. Cornils, *J. Mol. Catal. A: Chem.*, 1999, **143**, 1-10.
8. W. A. Herrmann, C. W. Kohlpaintner, *Angew. Chem., Int. Ed. Engl.*, 1993, **32**, 1524-1544.
9. M. Guerrero, A. Roucoux, A. Denicourt-Nowicki, H. Bricout, E. Monflier, V. Collière, K. Fajerwerg, K. Philippot, *Catalysis Today*, 2012, **183**, 34-41.
10. F. Hapiot, A. Ponchel, S. Tilloy, E. Monflier, *C. R. Chimie*, 2011, **14**, 149-166.
11. N. T. T. Chau, S. Handjani, J.-P. Guegan, M. Guerrero, E. Monflier, K. Philippot, A. Denicourt-Nowicki, A. Roucoux,*Chem.Cat.Chem.*, 2013, **5**, 1497-1503.
12. M. Ferreira, H. Bricout, A. Sayede, A. Ponchel, S. Fourmentin, S. Tilloy, E. Monflier, *Adv. Synth. Catal.*, 2008, **350**, 609-618.
13. M. Guerrero, Y. Coppel, Nguyet Trang Thanh Chau, A. Roucoux, A. Denicourt-Nowicki, E. Monflier, H. Bricout, P. Lecante, K. Philippot, *ChemCatChem*, 2013, **12**, 3802-3811.
14. S. Tilloy, G. Crowyn, E. Monflier, P.W.N.M. van Leeuwen, J.N.H. Reek, *New. J. Chem.* 2006, **30**, 377-383.
15. M. Ferreira, F.X. Legrand, C. Machut, H. Bricout, S. Tilloy, E. Monflier, *Dalton Trans.*, 2012, **41**, 8643-8647.

AUTHOR INDEX

SUBJECT INDEX

Printed in the United States
by Baker & Taylor Publisher Services

Printed in the United States
by Baker & Taylor Publisher Services